Evaporation—Combustion of Fuels

Joseph T. Zung, EDITOR

University of Missouri—Rolla

A symposium sponsored by
the Division of Petroleum
Chemistry, Inc. at the
172nd Meeting of the
American Chemical Society
San Francisco, Calif.,
August 31–September 1, 1976.

ADVANCES IN CHEMISTRY SERIES **166**

AMERICAN CHEMICAL SOCIETY
WASHINGTON, D. C.   1978

Library of Congress CIP Data

Evaporation-combustion of fuels.
  (Advances in chemistry series; 166 ISSN 0065-2393)

  Includes bibliographical references and index.

  1. Combustion engineering—Congresses. 2. Evapo-
ration—Congresses. 3. Fuel—Congresses.
  I. American Chemical Society. Division of Petroleum
Chemistry. II. Series.

QD1.A355 no. 166   [TJ254.5]   540'.8s
                                    [621.4'023] 78-9154
  ISBN 0-8412-0383-0   ADCSAJ   166 1-296 1978

Advances in Chemistry Series

Robert F. Gould, *Editor*

FOREWORD

ADVANCES IN CHEMISTRY SERIES was founded in 1949 by the American Chemical Society as an outlet for symposia and collections of data in special areas of topical interest that could not be accommodated in the Society's journals. It provides a medium for symposia that would otherwise be fragmented, their papers distributed among several journals or not published at all. Papers are reviewed critically according to ACS editorial standards and receive the careful attention and processing characteristic of ACS publications. Volumes in the ADVANCES IN CHEMISTRY SERIES maintain the integrity of the symposia on which they are based; however, verbatim reproductions of previously published papers are not accepted. Papers may include reports of research as well as reviews since symposia may embrace both types of presentation.

CONTENTS

PREFACE

Hydrocarbon combustion comprised at least 95% of the entire energy production in 1971. Reliable projections into the year 1985 estimate that, even with nuclear power and solar and geothermal energy playing a more prominent role, hydrocarbon combustion will still constitute about 85% of the energy sector. Thus for many years to come, research on combustion of fossil fuels should play an important role in our attempts to reduce energy consumption and to limit pollution.

Over the past two decades, the field of combustion research has been developed mainly by aerospace concerns and automotive engineers. The current energy dilemma, however, has led research scientists to redirect their effort to help the nation in search of energy independence by trying to achieve greater combustion efficiency, more effective energy conservation, and resource recovery. In spite of all the impressive scientific effort of the past, we still face enormous gaps in our knowledge of fuel combustion. For example, the most important and useful factor in energy conversion by combustion lies in the exothermic character of the combustion process; yet, paradoxically, of all the aspects of combustion, it is the chemistry of the actual evolution of energy that we understand the least.

Furthermore, the new environmental constraints and the associated lack of supply of clean fuels demand efficient power extraction without harmful effluents. This new dimension emphasizes further the need for closer collaboration between chemists and chemical engineers with other combustion scientists in our search for solutions to the nation's energy problems. Collaboration requires first a good line of communication between chemists and combustion scientists to make each other aware of the amount of knowledge already accumulated and of the types of problems to be solved. It is in fact appalling to notice the poor communications in the past, not only between chemists and the combustion engineering community, but sometimes even between workers in different but related fields, such as between aerospace and automotive engineers. Also needed is the awareness that basic research can assist at many different levels, from providing a framework at a quantitative level all the way to providing quantitative solutions to particular problems.

This symposium was organized to bridge the communication gap between combustion scientists and the chemical profession. It is the first time since 1956 that fuel spray combustion has been explored more or less exhaustively by a group of experts at an ACS national meeting.

The goal of the symposium is to provide a forum for chemists, chemical engineers, and combustion scientists to discuss the most recent advances in combustion research of fuel sprays in order to improve the efficiency in energy conversion, to eliminate soot and pollution by-products, and to improve the fuel quality by chemical means. It is also befitting the occasion of our Nation's Bicentennial Celebration and of our Society's Centennial Meeting to address ourselves to one of the most urgent problems of our nation—conservation of energy and environmental resources.

I would like to express my sincere thanks to the Petroleum Research Fund of the American Chemical Society for a grant to support this symposium and to the Division of Petroleum Chemistry, Inc. of the American Chemical Society for its generous sponsorship. I am indebted to all the authors and invited speakers who have generously accepted my invitation to contribute their knowledge and to share their precious time with us. The success of the symposium has been entirely due to them.

University of Missouri—Rolla
Rolla, MO 65401
August 31, 1976

JOSEPH T. ZUNG

Theory

Transient Heating and Liquid-Phase Mass Diffusion in Fuel Droplet Vaporization

W. A. SIRIGNANO

Princeton University, Princeton, NJ 08540

C. K. LAW

Northwestern University, Evanston, IL 60201

Theoretical and experimental advances on the transient, convective, multicomponent droplet vaporization are reviewed, with particular emphasis on the internal heat, mass, and momentum transport processes and their effects on the bulk vaporization characteristics. Whereas these processes have only small quantitative effects on single-component droplet vaporization, they can modify qualitatively the vaporization behavior of a multicompoent droplet. Depending on the intensity of convective transport through internal circulation, liquid-phase mass diffusion or the volatility differentials among the various components can become the rate-limiting factor. The generation of internal circulation through the surface shear induced by external gas streams is discussed. Some potentially interesting and practically important research problems are identified.

In many chemical power plants the fuel is introduced in the form of a spray. The droplets subsequently vaporize as they are dispersed in the reactor. Frequently chemical reactions are initiated prior to the complete vaporization of the droplets, and the flame will then propagate through a mixture consisting of fuel vapor, fuel droplets, and other gaseous components, including the oxidizer. Both direct and indirect studies on the performance of various types of engines indicate that the resulting combustion characteristics depend significantly on the amount and the sizes of the fuel droplets that are present when active chemical reactions are initiated. Hence studies on spray vaporization and combustion are of

0-8412-0383-0/78/33-166-003$06.50/0 © 1978 American Chemical Society

primary importance in analyzing and improving the performance of engines using spray injections.

Some simplified analyses have been performed for the combustion (1–9) and pure vaporization (10, 11, 12) of sprays. Recently more elaborate numerical codes on the unsteady, two-dimensional, two-phase, chemically reacting flows are also being developed (13). In all these studies the spray is frequently assumed to be sufficiently dilute such that the mutual interferences between the motion and the vaporization processes of the individual droplets are either completely neglected or are manifested only through their collective modifications of the state of the bulk gas. (The term vaporization is used here to imply both combustion and pure vaporization.) Hence the vaporization and kinematic behavior of a single, isolated droplet in an infinite expanse of gas serve as fundamental inputs to the spray analysis. These single-droplet phenomena are discussed in this chapter.

In certain practical situations the validity of the above dilute spray assumption may be substantially weakened. This is particularly serious for spray combustion exhibiting individual droplet combustion modes. The enveloping flames are close to each other and frequently may even overlap. In fact Chigier and his co-workers (14, 15) have asserted repeatedly from their experimental observations that the individual droplet combustion mode cannot exist in the oxygen-starved environment within a spray. Only limited research has been conducted on the combustion of closely spaced droplets (16, 17, 18, 19, 20).

The idealized problem in which we are interested can be stated as follows. At time $t = 0$ a droplet of known size, temperature, and

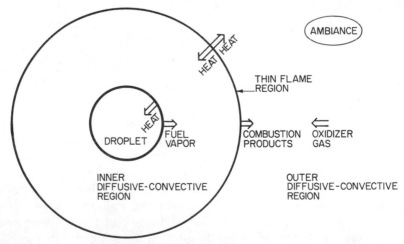

Figure 1a. *Flow configuration for spherically symmetric droplet vaporization*

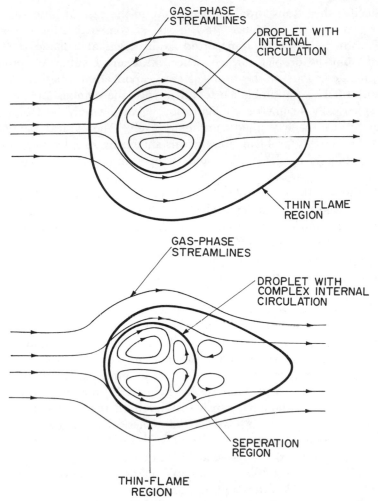

Figure 1b. (top) *Flow configuration for convective droplet vaporization without separation. Figure 1c.* (bottom) *Flow configuration for convective droplet vaporization with separation*

composition is injected, after ignition in the case of combustion, into a gas stream of infinite extent and of known flow velocity, temperature, and composition. In the droplet-stationary reference frame the several possible flow configurations are shown in Figure 1.

Figure 1a indicates an idealized, spherically symmetric situation when there is no relative velocity between the droplet and the gas stream. The only spacial dependence is the radial distance r. At the droplet surface the inwardly directed heat flux from the gas phase is used to effect vaporization as well as to heat the droplet interior. The fuel vapor produced is then transported outward and, in the case of pure vaporization, mixes with the ambiance. In the case of combustion the fuel vapor reacts with the inwardly transported oxidizer gas at a flame zone, detached from the droplet surface, where heat and combustion products are generated. Part of the heat sustains vaporization whereas the rest is transported to the ambiance together with the combustion products. This spherically symmetric configuration, involving only heat and mass transports, has been studied most extensively. The typical parameters of interest in this problem are the instantaneous droplet vaporization rate, droplet size, flame size and temperature, and the generated pollutants.

In Figures 1b and 1c relative motion exists between the droplet and the gas stream such that internal circulation is established through the shear stress induced at the nonrigid interface. Nonradial transports are important in the present cases, particularly for the momentum transport. Separation and reverse flow may (Figure 1c) or may not (Figure 1b) occur, although the analysis is expected to be greatly complicated for the former case.

Hence a complete analysis of the phenomena of interest will involve the simultaneous descriptions of the chemical reactions in the gas phase, the phase change processes at the interface, the heat, mass, and momentum transport processes in both the gas and liquid phases, and the coupling between them at the interface. The processes are transient and can be one dimensional (spherically symmetric) or two dimensional (axisymmetric). Although extensive research on this problem has been performed, most of it emphasizes the spherical-symmetric, gas-phase transport processes for the vaporization of single-component droplets. Fuchs' book (84) provides a good introduction to droplet vaporization whereas Wise and Agoston (21), and Williams (22), have reviewed the state of art to the mid-fifties and the early seventies, respectively.

This chapter complements Refs. 21 and 22 in reviewing the progresses made on the transient, convective, multicomponent droplet vaporization, with particular emphasis on the internal transport processes and their influences on the bulk vaporization characteristics. The interest and importance in stressing these particular features of droplet vaporization arise from the fact that most of the practical fuels used are blends of many chemical compounds with widely different chemical and physical properties. The approximation of such a complex mixture by a single compound, as is frequently assumed, not only may result in grossly inaccurate estimates of the quantitative vaporization characteristics but also may not account for such potentially important phenomena as soot formation when the droplet becomes more concentrated with high-boiling point compounds towards the end of its lifetime. Furthermore, multicomponent droplet vaporization is intrinsically transient in nature because of the possible continuous variations in the droplet composition. These variations, as will be shown, depend critically on the detailed descriptions of the convective transport mechanisms.

In the next section some of the important time scales are identified and transient droplet heating effects during the spherically symmetric, single-component droplet vaporization are reviewed. Spherically symmetric, multicomponent droplet vaporization and droplet vaporization with nonradial convection are discussed in later sections.

Transient Droplet Heating during Spherically Symmetric Single-Component Droplet Vaporization

General Discussions. The vaporization configuration is shown in Figure 1a. For pure vaporization the "flame" is simply the ambience. We first present some general concepts and identify certain important time scales applicable to many situations pertaining to droplet vaporization.

Since we are not concerned presently with ignition and extinction phenomena (*23, 24, 25, 26, 27*) caused by slow chemical heat release rates, the gas-phase chemical reactions can be assumed to occur at rates much faster than the gas-phase heat and mass transfer rates. This implies that the gas-phase consists of two (one in the case of pure vaporization) convective–diffusive regions separated by a flame of infinitesimal thickness, at which the outwardly diffusing fuel vapor reacts stoichiometrically and completely with the inwardly diffusing oxidizer gas.

By further assuming, realistically, that the interfacial phase-change rates are also much faster than the gas-phase processes, then the fuel vapor at the droplet surface is at its saturation value corresponding to the prevailing pressure and the temperature and liquid composition at the droplet surface.

For ambient pressures sufficiently less than the critical pressure of the fuel, the droplet remains in the liquid phase throughout its lifetime. The large liquid-to-gas density ratio then implies that the liquid droplet possesses large thermal and mass inertia compared with the gas phase; subsequently the gas-phase processes can be assumed to be quasi-steady. This assumption has been found to be very accurate even at moderate pressures (*28*). For near-critical or super-critical vaporization occurring typically in rocket motors and diesel engines, unsteady gas-phase analyses are required (*29–34, 85*).

For the present spherically symmetric configuration, from the low-speed and nonviscous nature of the flow, the radial pressure variations are negligible (*35*). Hence the momentum equation is simply represented by the statement that the pressure is uniform. This isobaric assumption has again been amply substantiated (28).

Since a planar flame is absolutely unstable to small disturbances (86, 87, 88), the small radii of curvature usually associated with droplet flames in general tend to stabilize these flames. Indeed, available experimental observations on droplet combustion do seem to indicate that they are stable.

Finally, to facilitate the mathematical developments, it is convenient to assume for the gas phase that: the second-order Soret and Dufour diffusion processes are unimportant; the heat capacities at constant pressure C_p, and the thermal conductivity coefficients k, are constants; the binary diffusion coefficients D are equal for all pairs of species; and the thermal and mass diffusion rates are equal such that the Lewis number is unity.

With the above assumptions, the spherically symmetric gas-phase heat and mass transport processes are described by the following system of first-order ordinary differential equations.

Continuity:

$$dm/dr = 0 \qquad (1)$$

Species:

$$\dot{m}Y_F - 4\pi\rho Dr^2 dY_F/dr = \dot{m} \qquad\qquad r_s < r < r_f \quad (2)$$

$$\dot{m}Y_O - 4\pi\rho Dr^2 dY_O/dr = -\dot{m}/v \qquad\qquad r_f < r < \infty \quad (3)$$

Energy:

$$\dot{m}C_p(T - T_s) - 4\pi kr^2 dT/dr = -\dot{m}H \qquad\qquad r_s < r < r_f \quad (4)$$

$$\dot{m}C_p(T - T_s) - 4\pi kr^2 dT/dr = -\dot{m}(H - Q) \qquad r_f < r < \infty \quad (5)$$

where $\dot{m} = 4\pi r^2 \rho v$ is the mass evaporation rate; ρ, v, Y, T, v, and Q are the density, the bulk gas velocity, the mass fraction, the temperature, the stoichiometric fuel-oxidizer mass ratio, and the chemical heat release per unit mass of fuel consumed, respectively; and the subscripts g, l, F, O, s, f, and ∞ will respectively designate conditions for the gas phase, the liquid phase, the fuel, the oxidizer, the droplet surface, the flame, and infinity. The function H is an effective specific latent heat of vaporization which consists of the specific latent heat of vaporization, L, plus the sensible heat required to heat the droplet interior per unit mass of fuel vaporized. The precise functional form of H then depends on the modeling of internal heat transfer and hence provides the coupling between the gas- and the liquid-phase processes.

Equations 2, 3, 4, and 5 can be readily integrated (*36*) to yield the various quantities of interest:

$$\beta = \ln\{1 + [C_p(T_\infty - T_s) + vY_{0\infty}Q]/H\} \tag{6}$$

$$\gamma = \beta/\ln(1 + vY_{0\infty}) \tag{7}$$

$$T_f = \{T_\infty + vY_{0\infty}[T_s + (Q - H)/C_p]\}/(1 + vY_{0\infty}) \tag{8}$$

and

$$H = \frac{(1 - Y_{Fs})[C_p(T_\infty - T_s) + vY_{0\infty}Q]}{Y_{Fs} + vY_{0\infty} - Y_{Ff}(1 + vY_{0\infty})} \tag{9}$$

where $\beta = \dot{m}/(4\pi\rho Dr_s) = -[(\rho_l C_p)/(2k)]dr_s^2/dt$ is a parameter representing the surface regression rate, $\gamma = r_f/r_s$ is the flame-front standoff ratio, and the fuel vapor concentration at the droplet surface, Y_{Fs}, is related to the droplet surface temperature through the Clausius–Clapeyron relation:

$$Y_{Fs} = Y_{Fs}(T_s) \tag{10}$$

The solutions are presented in such a way that they apply to pure vaporization by setting $Y_{0\infty} = 0$ and $Y_{Ff} = Y_{F\infty}$ and to combustion by setting $Y_{Ff} = 0$.

The discussions so far are quite general and hence are applicable to all cases involving the spherically symmetric vaporization of a single-component droplet. Equations 6, 7, 8, 9, and 10 show that for a given fuel oxidizer system and for prescribed ambient conditions $Y_{0\infty}$ and T_∞, the solutions are determined to within one unknown, H. Three models with different internal heat transport descriptions are presented below.

The d²-Law Model. Significant physical insight into the vaporization processes can be gained through this simplified model formulated by Spalding (*37*), Godsave (*38*), Goldsmith and Penner (*39*), and Wise et al. (*40*). The model assumes the droplet temperature to be uniform and constant throughout its lifetime, hence the heat conducted inward to the droplet surface is used only for vaporization, viz:

$$H = L \tag{11}$$

Subsequently the droplet surface area regression rate β, the flame-front standoff ratio γ, and the flame temperature T_f are all constants of the system and independent of the instantaneous droplet size. For combustion, since the computed T_s is usually close to the liquid boiling point T_b, Equations 9 and 10 are frequently replaced by the simple statement that $T_s = T_b$.

Despite its simplicity, the d²-law contains much information on the physical phenomena, and hence its results have been used extensively either to provide rough estimates for the droplet vaporization rate or to serve as input to the formulation of more complex processes, for example spray analyses.

The d²-law assumes a constant T_s. However, in many practical situations the temperature of the droplet when introduced into the evaporator is far below this final, equilibrium value. Hence an initial transient heating period exists during which β, γ, and T_f all increase whereas H decreases. Furthermore it can be estimated also that the sensible heat required to heat the droplet is of the same order as the latent heat of vaporization. Hence droplet transient heating effects on the bulk vaporization characteristics are expected to be significant. Two such models, representing extreme rates of internal heating, will be discussed.

The Diffusion Limit Model. In this model (*34, 41–45*) internal motion does not exist such that diffusion is the only heat transport mechanism. This represents the slowest internal heat transfer limit and is relevant for more viscous fuels or during the initial period when insufficient amount of internal motion has been generated.

The liquid-phase temperature variation is then governed by the heat conduction equation:

$$\frac{\partial T}{\partial t} = \frac{\alpha_l}{r^2} \frac{\partial}{\partial r} \left(r^2 \frac{\partial T}{\partial r} \right) \qquad 0 \leq r \leq r_s \qquad (12)$$

with the initial and boundary conditions:

$$T(r, 0) = T_o(r), \, T(r_s(t), t) = T_s(t), \, (\partial T / \partial r)_{r=0} = 0 \qquad (13)$$

where $T_o(r)$ is given, α_l is the liquid thermal diffusivity, and the subscript o designates the initial state. The parameter H is now given by:

$$H = L + \dot{m}^{-1} (4\pi r^2 k_l \partial T / \partial r)_{r=r_s} \qquad (14)$$

An analytical solution to Equations 12 and 13, subject to the coupling relation Equation 14, is difficult to obtain since one of the boundaries, viz. the droplet surface, is continuously regressing. Hence numerical integration was used to obtain the following results (*45*) on the combustion of an *n*-octane droplet in the standard atmosphere, with $T_o(r) = 300°K$. The total burning time is $0.236 r_{so}^2 / \alpha_l$.

Figure 2 shows the spacial distribution of the droplet temperature at different normalized times, t/t_{total}, where t_{total} is the total vaporization time. The bracketed term in Figure 2 is the fractional amount of fuel that has vaporized, which sometimes is a more meaningful measure for the duration of a phenomenon during droplet vaporization. After vaporization is initiated, the thermal wave reaches the droplet center only after a substantial time lag. A uniform temperature profile is not achieved until near burn-out.

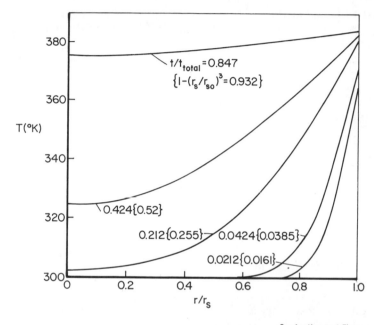

Combustion and Flame

Figure 2.   Temporal variations of the droplet temperature profiles for single-component droplet vaporization (45)

Figure 3 shows that the period with $\tilde{H} = H/L > 2$, which indicates that droplet heating is the dominant heat utilization mode at the droplet surface, occurs in the initial 10% of the droplet lifetime. After this period droplet vaporization becomes important, with $\tilde{H} < 2$. Comparing Figures 2 and 3, the system is already droplet-vaporization dominated ($\tilde{H} < 2$) before the central core is appreciably heated, implying that heating of the central core, which consists of only a small fraction of the droplet mass, does not require much heat and that vaporization proceeds efficiently once the surface layer is heated. It is also seen from Figure 3 that β, and consequently γ which is linearly proportional to β, can increase by an order of magnitude from the initial to the final state.

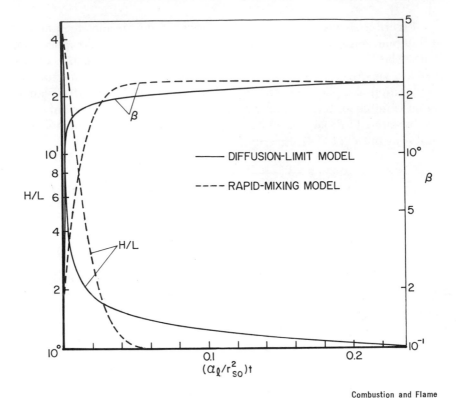

Combustion and Flame

Figure 3.   Temporal variations of the effective latent heat of vaporization
H/L *and the droplet surface regression rate parameter* β *with extreme inter-*
nal heat transport rates (45)

Figure 4 shows that after the initial 10–20% of the droplet lifetime, the droplet surface area appears to regress quite linearly with time. Hence if one is only interested in approximate estimates for the droplet size, then overall quasi-steadiness can be considered to be attained in about 20% of the droplet lifetime. However, if detailed droplet temperature variation is needed, then one would consider that unsteadiness prevails throughout the droplet lifetime.

The Rapid Mixing Model. In this model (*36, 46*) it is assumed that the combined internal convective–diffusive transports are so rapid that the droplet temperature is maintained spacially uniform but temporally varying. This represents the fastest internal heat transfer limit and is relevant for less viscous fuels in which internal motion can be easily generated. The model is somewhat artificial in that the existence of nonradial convection, particularly the requirement for the external gas stream to generate internal circulation, invalidates the assumption of spherical symmetry. However, by comparing results from the present model and the diffusion limit model, effects caused by the extremes in internal heat transport rates can be isolated without being complicated by additional effects from the differences in gas-phase transport modes. Results from these two models should provide lower and upper bounds on the expected evaporation behavior.

The parameter H is now given by:

$$H = L - C_{pl}dT_s/d(\ln r_s^3) \qquad (15)$$

Combining Equations 9, 10, and 15, the variation of r_s with T_s can be explicitly expressed.

When results of the rapid mixing model and the diffusion limit model are compared, it is shown (Figures 3 and 4) that in the initial droplet-heating dominated period, the former predicts a larger droplet size, a lower surface temperature, a larger H, a slower vaporization rate β, and a small flame size γ, as would be expected since more heat is needed to heat the entire droplet. The reverse situation prevails in the subsequent vaporization-dominated period.

Figure 4 also shows that the predicted temporal variations on the droplet surface area agree quite closely with each other, implying that

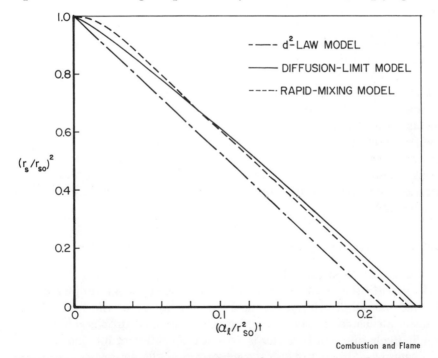

Combustion and Flame

Figure 4.   Temporal variations of the droplet surface area with extreme internal heat transport rates (45)

the droplet size variations, and subsequently the droplet vaporization time, can be predicted with good accuracy regardless of the model of internal heat transfer. This occurs since the surface layer is initially heated at approximately similar rates, hence heating of the much lighter inner core constitutes only small perturbations to the total heat budget at the surface, whether it takes place simultaneously with, or subsequent to, the surface heating process.

Finally, except for the initial period, the d^2-law is actually quite adequate, and hence because of its simplicity remains a strong competitor of the transient heating models in providing rough estimates for the droplet size and the total vaporization time.

Experimental Observations. Whereas many experiments have been performed on single-droplet vaporization (21, 22), most of them are conducted under the influence of either natural and/or forced convection, which not only distorts spherical symmetry, but also produces unwanted temporally varying convective effects on the vaporization process as the droplet size diminishes. The observations on the droplet surface area variations, however, do agree qualitatively with the predicted behavior of the transient models shown in Figure 4.

The only convection-free experiments are those of Kumagai and his co-workers (47, 48), in which a free droplet undergoes combustion in a freely falling chamber. The observed behavior on the temporal variations of the droplet and flame sizes agreed very well with predictions from the transient analyses (36, 45).

Another experiment of interest is that by El Wakil et al. (49), in which the center and peripheral temperatures were measured for a vaporizing droplet subjected to mild forced convection. The measurements show that there are essentially no differences between these two temperatures, not even during the initial transient heating period. Visual observations also revealed the existence of fairly rapid internal circulations. These imply that the assumption of a uniform droplet temperature may be quite realistic for droplet vaporization with some external convective motion.

Spherically Symmetric Multicomponent Droplet Vaporization

General Discussion. It was shown in the previous section that the bulk vaporization characteristics of a single-component droplet do not depend too sensitively on the detailed description of the internal heat transfer mechanisms. However, for multicomponent droplet vaporization qualitatively different behavior is expected for different internal transport mechanisms. This is because the vaporization characteristics (for example, the vaporization rate, the flame temperature and location, and the nature and amount of pollutants generated) depend directly on the relative amounts of different species that are vaporized at the surface and are subsequently transported to the flame. These in turn are controlled by the volatility differentials among the various components and by the rates with which they can be transported from the droplet interior to the surface. Obviously the more volatile species that cannot be rapidly transported to the surface will not contribute substantially to the amount of fuel vaporized.

The "General Discussion" of the previous section is equally applicable here, except now proper multicomponent descriptions of the gasphase transport and the interfacial phase change should be used (*50, 51, 52*). By assuming the gas-phase reactions are again confined to a flamesheet where the reactants are consumed in a species-weighted stoichiometric proportion, explicit expressions can be derived (*50*) for β, γ, T_t, H, and the fractional mass evaporation rate of the ith species, as functions of the temperature and vapor concentration at the droplet surface.

The vapor concentration of the ith species at the surface is given by:

$$Y_{Fs+} = Y_{Fs+}(a_i, Y_{Fs-}, T_s) \qquad (16)$$

where a_i is the activity coefficient. If the mixture can be assumed to be ideal, then $a_i = 1$, and Raoult's law holds. This is the simplest case conceivable (*53*), from both mathematical and physical view-points, and in fact approximately holds for most hydrocarbon fuels (*54*).

When a_i deviates significantly from unity, as would be the case when not all of the compounds are completely miscible, then separate computations for a_i are required. This can be accomplished by using some of the recently established models (*55*) which compute the a_i's by using group interaction parameters obtained from data reduction.

In the following sections we shall again investigate separately the vaporization characteristics for the diffusion limit and the rapid mixing limit.

The Diffusion Limit Model. In this model heat and mass in the droplet are transported by diffusion alone. Since liquid-phase mass diffusivities are typically of the order of 10^{-5} cm²/sec, which, when compared with typical droplet surface regression rates of 10^{-3} cm²/sec for combustion and 10^{-4}–10^{-5} cm²/sec for pure vaporization, shows that except for very slow rates of vaporization or very small droplet sizes (*89*), liquidphase mass diffusion is the rate-limiting process. Hence most of the more volatile compounds will be trapped within the droplet interior, unable to reach the droplet surface to become preferentially vaporized. Further realizing that the liquid-phase thermal diffusivities are of the order of 10^{-3} cm²/sec, the droplet interior will be heated substantially during most of its lifetime. Hence the compounds in the interior may experience temperatures exceeding their local boiling points, causing the droplet to become unstable.

The above order-of-magnitude arguments are substantiated (56, 89) for a heptane–octane droplet vaporizing in a 2300°K, 1 atm, environment. Figure 5 shows the mass fraction of heptane at the center and the surface. Mass diffusion is so slow that the center value is practically unperturbed during the entire droplet lifetime. The surface value initially decreases during a short transient period and then attains a constant value. This constant behavior prevails until towards the end of the droplet lifetime when the droplet size has become comparable with the characteristic diffusion length. Then the volatility differentials again become the rate-limiting factor such that heptane, being more volatile, is quickly depleted from the droplet composition. Hence one consequence of the slow mass diffusion is that the droplet concentration profile, and therefore its composition and fractional mass evaporation rate, are constants during most of the droplet lifetime.

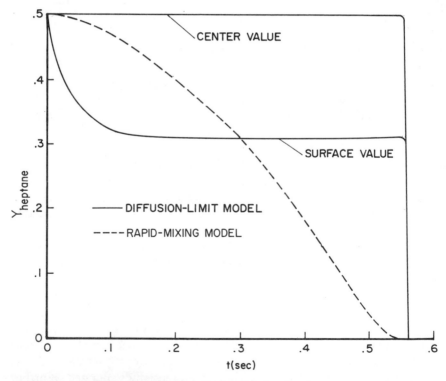

Figure 5. Temporal variations of the droplet heptane composition for a heptane–octane droplet, with extreme internal heat and mass transport rates (56, 89)

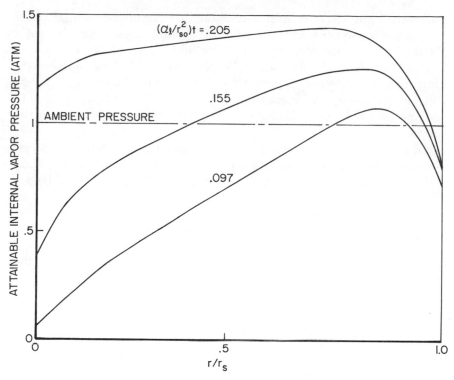

Figure 6.   Attainable internal vapor pressure for a heptane–octane droplet with internal diffusive transport (56)

Figure 6 shows that the hypothetical vapor pressure attainable within the droplet can be quite substantial and can easily exceed the external pressure. Hence vapor bubbles can be formed within the droplet through either heterogeneous or homogeneous nucleation. For sufficiently intense internal pressure build-up, the surface tension force can be overcome, causing the parent droplet to rupture. This micro-explosion phenomenon also occurs during the combustion of fuel droplets emulsified with water or methanol (57, 58, 59, 90). The rupturing of the parent droplet dispenses a mist of much finer micro-droplets in the gas stream, hence creating a more uniform charge for combustion. The amount of heterogeneous droplet combustion is also reduced because of the smaller droplet sizes and hence may lead to significant improvements in the combustion characteristics from energy efficiency and pollutant reduction considerations.

The Rapid Mixing Model. In this model (50, 56, 60, 61) the concentration and temperature within the droplet are maintained spacially uniform but temporally varying. Since the droplet surface is continuously replenished with the more volatile compounds, volatility differential rather than liquid-phase mass diffusion is the rate-limiting factor. Consequently Figure 5 shows that the composition of the more volatile compound, heptane, continuously decreases. For increased volatility differentials, either from the inherent property differences or when vaporization is less efficient, the more volatile compounds decrease at even faster rates and become depleted from the droplet composition at an early stage in the droplet lifetime.

In general, results from this model show (50) that the dominant vaporizing species participate approximately sequentially in order of their relative volatilities. The transition between any adjacent pair begins when the concentration of the less volatile species exceeds that of the more volatile one. The droplet temperature increases monotonically, with most of the increase occurring during the transition period, which can be quite abrupt for compounds with large volatility differentials. These vaporization characteristics differ qualitatively from those of the diffusion limit model.

Experimental Observations. Most of the experimental observations on multicomponent droplet vaporization use two-component droplets (49, 62, 63, 64). The observations on the temporal behavior of the droplet temperature (49, 64), size (62, 63), and composition (64) all indicate that the vaporization processes are controlled by the volatility differentials rather than by liquid-phase mass diffusion. Since the fuels used in these experiments are quite nonviscous, the above results then indicate that internal circulation of sufficient strength has been generated by the prevailing forced and/or natural convection.

Explosive combustion has been observed also (57, 58, 59, 62) for heavy oil droplets, with or without water emulsification, suspended on quartz fibers, indicating that internal boiling can occur. It is suspected (59), however, that the boiling is induced by heterogeneous nucleation at the suspension fiber and hence bears little relevance to the practical situation of free droplet vaporizing in a spray. Recently, the explosion of a freely falling emulsified droplet, ignited by flash-lamp discharge, has been observed (65), although the explosion could have been induced by the initial intense radiative loading. An unambiguous emulsified droplet explosion induced by thermal heating has yet to be observed.

Convective Multicomponent Droplet Vaporization

General Discussion. We have shown that for the vaporization of practical, multicomponent droplets, qualitatively different vaporization behavior results when extreme internal transport rates are assumed. Since diffusive transport is always present during the transient, it is the intensity of internal circulation that determines the resultant heat and mass transport rates. For a droplet translating in a gas stream, the primary mechanism through which internal circulation can be generated and maintained is through the shear stress induced at the nonrigid interface. This shear stress is in turn affected by the interfacial heat and mass transfer processes. The problem is a complex one; the two possible vaporization configurations are shown in Figures 1b and 1c.

The analysis can be significantly simplified by realizing that the rate with which the vorticity diffuses inwards, and hence establishes the fluid motion, is represented by the kinematic viscosity coefficient, which is of the order of 10^{-2} cm^2/sec and is at least one order of magnitude greater than the droplet surface regression rate. Hence quasi-steadiness for both the gas and liquid motion, with a stationary droplet surface and constant interfacial heat and mass flux, can be assumed. Once the fluid mechanical aspect of the problem is solved, the transient liquid-phase heat and mass transfer analyses, with a regressing droplet surface, can be performed.

A reasonably complete analysis for this problem has not been performed. However, many important contributions towards analyzing a particular aspect of the problem do exist and are discussed briefly according to the intensity of the external convection, which can be represented by the droplet Reynolds number Re. In all the discussions it is assumed that the droplet surface tension is sufficiently great to maintain its spherical shape.

Small Reynolds Number Flow, Re $<$ 1. The slow viscous motion without interfacial mass transfer is described by the Hadamard (*66*)–Rybcynski (*67*) solution. For infinite liquid viscosity the result specializes to that of the Stokes flow over a rigid sphere. An approximate transient analysis to establish the internal motion has been performed (*68*). Some simplified heat and mass transfer analyses (*69, 70*) using the Hadamard–Rybcynski solution to describe the flow field also exist. These results are usually obtained through numerical integration since analytical solutions are usually difficult to obtain.

The gas-phase slow viscous flow over a rigid sphere with interfacial mass injection or abstraction has been analyzed in Refs. *71, 72, 73,* and *74.* In general evaporation reduces the droplet drag significantly.

Intermediate Reynolds Number Flow, 1 $<$ Re $\underset{\sim}{<}$ 100. Hamielec and Johnson (*75*) obtained approximate expressions for the stream functions of both the inner and outer flows. Brounshtein et al. (*69*) used this result to analyze the heat and mass transfer for droplets and gas bubbles in liquid flows.

Large Reynolds Number Flow, Re $\gg$ 1. This is probably the more relevant limit for spray vaporization in many combustors involving high-pressure injection techniques. Simple estimates (52, 76) have shown that the Reynolds numbers associated with the droplet motion as well as the generated liquid motion remain quite large, typically above 100, over most of the droplet lifetime. These large Reynolds numbers imply that viscous effects in both the gas and liquid phases are confined to thin boundary layers across the interface.

In the absence of separation, the gas-phase free stream is described by the potential flow solution over a sphere (77):

$$\Psi_g = (1/2)u_\infty(1 - r_s^3/r^3)r^2\sin^2\theta \tag{17}$$

whereas the motion for the droplet inner core is given by the Hill's vortex (77):

$$\Psi_l = A(1 - r^2/r_s^2)r^2\sin^2\theta \tag{18}$$

with a strength to be determined by requiring that the vorticity discharged from the boundary layer at the rear stagnation point be conserved as it is convected forward, through the inviscid internal wake (along the axis of symmetry) into the boundary layer at the front stagnation point (78). In the above Ψ is the stream function, u_∞ the free stream gas velocity, θ the polar angle, and A the constant to be determined.

When the densities of the two fluids are of comparable magnitude, the velocities within the boundary layers do not differ significantly from their free stream values. Subsequently the governing equations can be linearized and solved (78, 79, 80, 81). This case is expected to be relevant for near-critical vaporization, provided the droplet can still maintain its spherical shape.

For a liquid droplet in a gaseous medium, the densities differ significantly. Whereas the linearized treatment can still be extended to the liquid-phase analysis, for the gas phase the conventional boundary layer analysis should be used (82). For high Reynolds number flow over a solid sphere, approximate solutions have been obtained using both the Blasius series and the momentum integral techniques (82).

The presence of adverse pressure gradients induces separation towards the rear stagnation point (Figure 1c). Whereas its influence on the gas-phase motion may not be critical, its effect on the liquid-phase motion may be quite significant because of the closed streamlines. If this is the case, then entirely different formulations may be required to account for the effects from separation and the wake. Such a detailed analysis has not been performed for either the gas or the liquid flow.

Strongly Convective Limit, The Flat Plate Model. It has been demonstrated that because of the complex geometrical configuration and the presence of pressure gradients, a detailed analysis for the convective droplet vaporization is extremely difficult. In order to gain insight into the effects of gas-phase convection on the liquid-phase heat, mass, and momentum transport, and realizing that the pressure gradient is very small along the flow directions at the shoulder region of the droplet, a flat-plate boundary layer-type analysis has been performed for the flow around this region (*51, 52, 76*). The flow configuration is shown in Figure 7 where a compressible gas stream flows over a parallel incompressible multicomponent liquid stream. Because of the self-similar nature of the flat-plate boundary layer flow, all the flow quantities are functions of a similarity variable η, which is proportional to $y/x^{1/2}$.

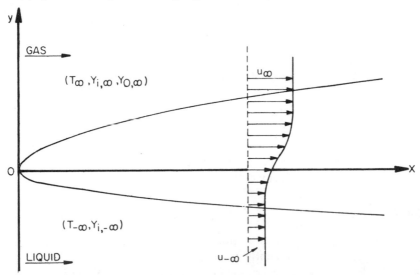

Figure 7.   Configuration of the flat-plate vaporization model

Figure 8 shows the x-velocity, concentration, and temperature profiles in the η-space for a gaseous stream with $Y_{O\infty} = 0.232$ and $T_\infty = 300°K$ flowing over stagnant hexane–octane mixture with $T_{-\infty} = 300°K$ and equal mass fractions. The induced surface velocity can be quite significant, in this case about 2% of the ambient gas velocity. This velocity increased with decreasing vaporization rates and liquid viscosities, the latter in particular. In the liquid the more/less volatile compounds respectively diffuse outward/inward, although the net mass fluxes for all species are directed outward. The liquid boundary layer thicknesses for heat, mass, and momentum transport differ significantly from each other, enabling simplifications to be made in a more detailed modeling

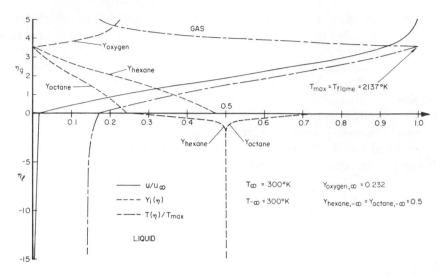

Combustion Institute

Figure 8. Profiles of x-velocity, concentration, and temperature for atmospheric air flowing over and reacting with a stagnant hexane–octane mixture (51, 52)

effort. Finally, the more volatile compounds are always preferentially vaporized, with the difference being more pronounced for high rates of liquid mass diffusion or under conditions unfavorable for vaporization. These agree with our previous observations.

Concluding Remarks

Droplet vaporization is a phenomenon occurring in a gas–liquid system, although only recently have serious efforts been made towards understanding the various liquid-phase processes and their influence on the overall behavior. The problem is a complex, yet interesting and important one. Fundamental research in the interdisciplinary areas of fluid mechanics, chemical kinetics, phase equilibrium analysis, and heat and mass transfer are required to achieve a good understanding of the problem. The following discussions may substantiate this point and stimulate future research efforts.

As discussed in the previous section, the convective droplet vaporization case has yet to be analyzed completely. The major difficulty lies with describing the fluid mechanical aspect of the phenomena, particularly for large Reynolds number flow when separation and reverse flow occur towards the rear stagnation point in both the gas and liquid phases. The difficulty is further compounded when the components in the droplet are not completely miscible, as is the case for emulsified fuels. The droplet then consists of more than one (liquid) phase, and hence multi-phase flow analyses are required even for the interior motion. It is also unclear, say for the emulsified fuels, how stable are the dispersed-phase micro-droplets when they collide into one another and as the parent droplet is continuously heated. Coalescence would immensely complicate the analysis.

A parallel experimental effort to study the internal motion may also demand design ingenuity. To be realistic, the droplets under testing are to be small (of the order of 100 μm), free from any suspension device, and are subjected to the shearing action of high-speed gas stream. Hence observation of stationary droplets may not be easy. Furthermore, direct optical probing of the droplet interior may also be difficult because of the severe degree of internal reflection for the small droplets involved.

For the combustion of high-boiling point mixtures, the droplet can become substantially heated. Subsequently pyrolysis of these compounds may occur and soot may eventually form. The mechanisms and rate constants of these liquid-phase reactions need be determined.

For emulsified fuel droplet combustion, the kinetic criteria for the onset of nucleation, and hence the possible rupturing of the parent drop-let, are required. If nucleation is likely to occur at the interface between the micro-droplets and the bulk liquid medium, then the influence of the emulsifying agents on the onset of nucleation should be assessed also. At the surface of the parent droplet, the vapor concentration depends on the accessibility of the molecules of the liquid phases to those of the gas phase. It is conceivable that the emulsifying agents and/or surface tension effects can prevent the dispersed liquid-phase micro-droplets from being in contact with the gas, hence effectively inhibiting its vaporization.

Finally, it is important to be reminded constantly of some of the basic assumptions adopted in describing the present class of droplet vaporization phenomena, particularly those of dilute spray and quasi-steady gas-proc-esses. These assumptions fall in certain practical situations; improvements in these aspects of modeling are essential.

Acknowledgements

This research was funded by the National Science Foundation under Grant No. NSF-RANN AER 75-09538. It was conducted while the second author was a Research Staff Member at Princeton University.

Literature Cited

1. Probert, R. P., *Philos. Mag.* (1946) **37**, 94.
2. Spalding, D. B., *Aero. Quart.* (1959) **10**, 1.

3. Williams, F. A., *Combust. Flame* (1959) **3**, 215.
4. Williams, F. A., Penner, S. S. Gill, G., Eckel, E. F., *Comb. Flame* (1959) **3**, 355.
5. Williams, F. A., *Phys. Fluids* (1958) **1**, 541.
6. Williams, F. A., *Symp. (Int) Combust.* (Proc.) *8th* (1962) 50.
7. Mizutani, Y., *Combust. Sci. Technol.* (1972) **6**, 11.
8. Polymeropoulous, C. E., *Combust. Sci. Technol.* (1974) **9**, 197.
9. Polymeropoulous, C. E., "Ignition and Propagation Rates for Flames in a Fuel Mist.," Report No. **FAA-RO-75-155**, U. S. Dept. of Transport, Washington, D. C., 1975.
10. Dickenson, D. R., Marshall, W. R., *AIChE J.* (1968) **14**, 541.
11. Law, C. K., *Int. J. Heat Mass Transfer* (1975) **18**, 1285.
12. Law, C. K., *Combust. Sci. Technol.* (1977) **15**, 65.
13. Bracco, F. V., Gupta, H. C., Krishnamurthy, L., Santavicca, D. A., Steinberger, R. L., Warshaw, V., *SAE Tech. Pap.* (1976) **760114**.
14. Nuruzzaman, A. S., Hedley, A. B., and Beer, J. M., *Symp. Combust., 13th* (1971) 787.
15. Chigier, N. A., *Symp. Combust. 15th* (1975) 453.
16. Rex, J. F., Fuhs, A. E., Penner, S. S., *Jet Propul.* (1956) **26**, 179.
17. Kanevsky, J., *Jet Propul.* (1956) **26**, 788.
18. Fedoseeva, N. V., "Advances in Aerosol Physics," V. A. Fedoseev, Ed., No. 1, 21; No. 2, 110; No. 3, 27; No. 3, 35, Wiley, New York, 1971.
19. Konyushenko, A. G., Fedoseeva, N. V., "Advances in Aerosol Physics," No. 7, p. 5, V. A. Feodseev, Ed., Wiley, New York, 1971.
20. Suzuki, T., Chiu, H. H., *Proc. Int. Symp. Space Technol. 9th* (1971).
21. Wise, H., Agoston, G. A., *Adv. Chem. Ser.* (1958) **20**, 116.
22. Williams, A., *Combust. Flame* (1973) **21**, 1.
23. Tanifa, C. S., Perez Del Notario, P., Moreno, F. G., *Sym. Combust. 8th* (1962) 1035.
24. Peskin, R. L., Wise, H., *AIAA J.* (1966) **4**, 1646.
25. Kassoy, D. R., Williams, F. A., *Phys. Fluids* (1968) **11**, 1343; (1969) **12**, 265.
26. Law, C. K., Williams, F. A., *Combust. Flame* (1972) **19**, 393.
27. Law, C. K., *Combust. Flame* (1975) **24**, 89.
28. Hubbard, G. L., Denny, V. E., Mills, A. F., *Int. J. Heat Mass Transfer* (1975) **18**, 1003.
29. Spalding, D. B., *ARS J.* (1959) **29**, 825.
30. Rosner, D. E., *AIAA J.* (1967) **5**, 163.
31. Matlosz, R. L., Leipziger, S., Torda, T. P., *Int. J. Heat Mass Transfer* (1972) **15**, 831.
32. Manrique, J. A., Borman, G. L., *Int. J. Heat Mass Transfer* (1969) **12**, 1081.
33. Crespo, A., Linan, A., *Combust. Sci. Technol.* (1975) **11**, 9.
34. Waldman, C. H., *Symp. Combust. 15th* (1975) 429.
35. Williams, F. A., "Combustion Theory," Reading, Mass, 1965.
36. Law, C. K., *Combust. Flame* (1976) **26**, 17.
37. Spalding, D. B., *Symp. Combust. 4th* (1953) 847.
38. Godsave, G. A. E., *Symp. Combust. 4th* (1953) 818.
39. Goldsmith, M., Penner, S. S., *Jet Propul.* (1954) **24**, 245.
40. Wise, H., Lorell, J., Wood, B. J., *Symp. Combust. 5th* (1955) 132.
41. Wise, H., Ablow, C. M., *J. Chem. Phys.* (1957) **27**, 389.
42. Parks, J. M., Ablow, C. M., Wise, H., *AIAA J.* (1966) **4**, 1032.
43. Waldman, C. H., Kau, C. J., Wilson, R. P., *West. Sec. Combust. Inst. Pap.* (1974) **74-17**.
44. Sangiovanni, J. J., Kestin, A. S., unpublished data, 1975.
45. Law, C. K., Sirignano, W. A., *Combust. Flame* (1977) **28**, 175.
46. Williams, F. A., *J. Chem. Phys.* (1960) **33**, 133.

47. Kumagai, S., Sakai, T., Okajima, S., *Symp. Combust. 13th* (1971) 779.
48. Okajima, S., and Kumagai, S., *Symp. Combust. 15th* (1975) 401.
49. El Wakil, M. M., Priem, R. J., Brikowski, H. J., Myers, P. S., Uyehara, O. A., *NACA* **TN-3490** (1956).
50. Law, C. K., *Combust. Flame* (1976) **26**, 219.
51. Law, C. K., Sirignano, W. A., 1975 Fall Meeting of the Eastern States Section/Combustion Institute, Paper No. 4, 1975.
52. Law, C. K., Prakash, S., Sirignano, W. A., *Symp. Combust. 16th* (1977) in press.
53. Guggenheim, E. A., "Mixtures," Oxford University, London, 1952.
54. King, M. B., "Phase Equilibrium in Mixtures," Pergamon, New York, 1969.
55. Fredenslund, A., Jones, R. L., Prausnitz, J. M., *AIChe J.* (1975) **21**, 1086.
56. Landis, R. B., Mills, A. F., *Proc. Int. Heat Transfer Conf. 5th,* (1974) No. **B7.9.**
57. Ivanov, V. M., Nefedov, P. I., *NASA TT* **F-258** (1965).
58. Dryer, F. L., unpublished data (1975).
59. Dryer, F. L., Rambach, G. D., Glassman, I., *Symp. Combust. 16th* (1977) in press.
60. Faeth, G. M., *AIAA Pap.* (1970) **70-7.**
61. Newbold, F. R., Amundson, N. R., *AIChe J.* (1973) **19**, 22.
62. Wood, B. J., Wise, H., Inami, S. H., *Combust. Flame* (1960) **4**, 235.
63. Kobayasi, K., "A Study on the Evaporation Velocity of a Single Liquid Droplet," *Technol. Rep. Tohoko Univ.,* XVIII, (2).
64. Glushkov, V. E., "Advances in Aerosol Physics," V. A. Fedoseev, Ed., No. 1, p. 12, Wiley, New York, 1971.
65. Spadaccini, L. J., "Evaluation of Dispersion Fuels," Report **R75-213717-1,** United Aircraft Research Laboratories, 1975.
66. Hadamard, J., *Compt. rend* (1911) **152**, 1735.
67. Rybcynski, W., *Bull. Acad. Cracovia* (1911) **A40.**
68. Savic, P., "Circulation and Distortion of Liquid Drops Falling Through a Viscous Medium," *Nat. Res. Counc. Can. Div. Mech. Eng. Lab. Tech. Rep.* (1953) **MT-22.**
69. Brounshtein, B. I., Zheleznyak, A. S., Fishbein, G. A., *Int. J. Heat Mass Transfer* (1970) **13**, 963.
70. Brunson, R. J., Wellek, R. M., *AIChe J.* (1971) **17**, 1123.
71. Fendell, F. E., Coats, D. E., Smith, E. B., *AIAA J.* (1968) **6**, 1953.
72. Golovin, A. M., *J. Eng. Phys.* (1973) **24**, 175.
73. Gal-Or, B., Yaron, I., *Phys. Fluids* (1973) **16**, 1826.
74. Muggia, A., *Aeroteknica* (1956) **36**, 127.
75. Hamielec, A. E., Johnson, A. I., *Can. J. Chem. Eng.* (1962) **40**, 41.
76. Prakash, S., Sirignano, W. A., unpublished data (1975).
77. Batchelor, G. K., *J. Fluid Mech.* (1956) **1**, 177.
78. Harper, J. F., Moore, D. W., *J. Fluid Mech.* (1968) **32**, 367.
79. Chao, B. T., *Phys. Fluids* (1962) **5**, 69.
80. Chao, B. T., *ASME Trans. Ser. C,* (1969) **91**, 273.
81. Parlange, J.-Y., *Acta Mech.* (1970) **9**, 323.
82. Schlichting, H., "Boundary Layer Theory," McGraw Hill, 1968.
83. Labowsky, M., Rosner, D. E., "Conditions for Group Combustion of Droplets in Fuel Clouds," *Adv. Chem. Ser.* (1977) **000** ..
84. Fuchs, N. A., "Evaporation and Droplet Growth in Gaseous Media," Pergamon, 1959.
85. Parlange, J.-Y., *Acta Mech.* (1973) **18**, 157.
86. Landau, L. D., *Acta Physicochem. URSS* (1944) **19**, 77.
87. Parlange, J.-Y., *J. Chem. Phys.* (1968) **48**, 1843.
88. Sivashinsky, G. I., *Int. J. Heat Mass Transfer* (1974) **17**, 1499.
89. Law, C. K., "Internal Boiling and Superheating in Vaporizing Multicomponent Droplets," *Acta Astronautica* (1977) in press.

90. Law, C. K., "A Model for the Combustion of Oil/Water Emulsion Droplets," *Combust. Sci. Technol.* (1977) in press.

RECEIVED December 29, 1976.

A Model for the Nonsteady Ignition and Combustion of a Fuel Droplet

J. J. SANGIOVANNI

United Technologies Research Center, East Hartford, CT 06108

Integral equation methods are used to analyze the nonsteady, spherically symmetric heating, ignition, and combustion of a single-component fuel droplet moving relative to an oxidizing ambient. Gas-phase heat and mass transfer are regarded as quasi-steady processes, and transient conduction is assumed to be the only transport mechanism within the droplet. The nonsteady nature of droplet temperature, flame temperature, burning rate, and flame position as affected by the mode of gas-phase heat and mass transfer are examined. Predictions immediately following ignition suggest that the flame surface rapidly collapses toward the droplet surface after ignition and before expanding away from the droplet during combustion. The effect of gas phase convection is to move the flame surface closer to the droplet and to decrease the flame temperature while the burning rate is essentially unaffected. The model agrees well with experimental observations for the temporal variation of droplet surface area and flame dimensions.

Since the earliest theoretical models by Spalding (1) and Godsave (2) describing the quasi-steady, spherically symmetric combustion of individual fuel droplets in quiescent atmospheres, numerous more elaborate theories have been proposed to provide a better understanding of droplet spray combustion. These theories are based on the premise that the physical and chemical processes involved during single-droplet combustion are fundamental to complex spray combustion processes.

Of the many extensions of the classical quasi-steady droplet combustion models, most have dealt with the nonsteady effects which take place during droplet ignition and combustion (3–9). Using Green's function techniques to evaluate the nonsteady heat and mass transfer equations for the infinite, spherically symmetric gas phase surrounding a stationary fuel droplet at the boiling temperature, Kesten and Sangiovanni (3) examined the temporal variation of flame structure and the resultant effect on nitric oxide production. They found that both flame temperature and flame position can vary significantly with time during the early stages of droplet combustion as a result of nonsteady gas-phase processes. Recently, Crespo and Linan (4) confirmed these effects of nonsteady gas-phase processes using an asymptotic analysis of the time-dependent gas-phase transport equations. Nonsteady effects resulting from droplet heating have been studied by Law (5) and Law and Sirignano (6) for droplet combustion and by Sangiovanni and Kesten (7) for self-ignition of droplets by assuming quasi-steady gas-phase processes. The former studies of droplet combustion (5, 6) have demonstrated that droplet heating contributes significantly to the nonsteady behavior observed by Okajima and Kumagi (8) during the initial period of droplet combustion. The most complete theoretical study of nonsteady droplet combustion was by Kotaki and Okazaki (9) who numerically integrated the time-dependent conservation equations for both the liquid and gas phases and predicted that significant nonsteady effects persist throughout the entire droplet combustion process. However, the accuracy of their numerical computations has been questioned as a result of a similar numerical solution reported by Hubbard et al. (10) for the case of droplet vaporization. Contrary to the findings of Kotake and Okazaki, Hubbard et al. showed that nonsteady effects exist for about the first 20–30% of the droplet lifetime, and gas-phase transport processes can be treated as quasi-steady for most practical cases of interest.

In practical liquid spray combustion devices, burning may occur either in a diffusion flame which envelops one or more droplets or in a wake flame behind the droplets depending on the droplet spray density and relative motion with respect to the combustion gas environment. If the droplets in the spray are extremely small, vaporization may occur prior to reaction of the fuel vapor and oxidizer in a partially premixed flame. From an experimental study of the combustion of monosized droplet streams in stationary self-supporting flames, Nuruzzaman et al. (11) noted significantly lower burning rate coefficients compared with single-droplet measurements which they ascribed to oxygen depletion as a result of droplet interaction. Chigier et al. (12) have asserted from their experimental observations that the burning of individual droplets cannot exist within the oxygen-depleted spray combustion flame. By assuming that droplets within a spray do not burn individually but instead vaporize within a turbulent diffusion flame, Onuma and Ogasawara (13) analyzed the spray combustion flame using vaporization rates for single droplets and obtained fairly good agreement with experimental observations. If the individual-droplet model is to be the basis for a more complex spray combustion model, the single-droplet model must take into account the effects of droplet relative motion which enhances the convection of heat and mass transfer with the hot gas environment. With this objective in mind, this chapter formulates and evaluates a mathematical model for describing the nonsteady heating, vaporization, and subsequent self-ignition and combustion of a single fuel droplet which is moving relative to a hot combustion gas stream.

Although several theoretical models have been proposed to evaluate the effect of relative motion on single-droplet combustion (*14, 15, 16*), none of these models accounts for the equally important nonsteady liquid-phase heating. Using an effective oxidizer diffusion coefficient based on an empirical Nusselt number correlation for droplet vaporization, Brzustowski and Natarajan (*14*) modified the classical quasi-steady droplet burning model to account for the influence of a convective field. A more detailed analysis has been proposed by Fendel et al. (*15*) by assuming Stokes flow together with a flame surface approximation to predict the mass burning rate for single-droplet combustion in a slow viscous flow. This model shows how convective transport as a result of relative motion enhances the pure diffusion-controlled combustion process. Isoda and Kumagi (*16*) developed the first analysis showing the influence of convection on flame shape by considering the droplet to be a point source of vapor and assuming Navier–Stokes flow.

Mathematical Model

Formulation and Governing Equations. The mathematical model presented here describes the nonsteady heating, vaporization, and subsequent self-ignition and combustion of a single-component fuel droplet which is moving relative to a hot combustion gas stream. Figure 1 is a schematic of the droplet combustion system. The system under consideration is assumed to be spherically symmetric, isobaric, and at low pressure (typically less then one-tenth of the fuel critical pressure). Conduction is assumed to be the only mode of heat transport within the droplet such that the droplet temperature can vary both spatially and temporally. Although the mathematical complexities imposed by this last assumption can be avoided if rapid circulation exists within the droplet of the strength postulated by El Wakil et al. (*17*) and Law (*5*), such circulation would be unlikely within a liquid droplet having either a small diameter or a high viscosity. Heat and mass transfer within the gas phase are considered to occur by conduction and diffusion as well as by the radial convective transport of mass resulting from droplet vaporization. Since movement of the droplet relative to the gas stream enhances the transfer of heat and

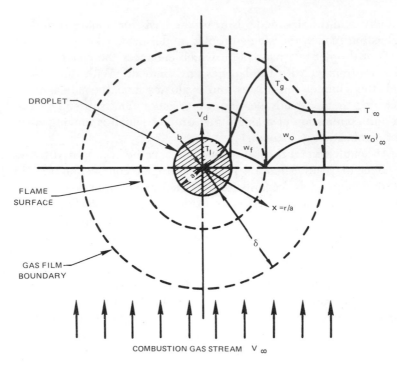

Figure 1.   Droplet combustion model

mass in the region immediately surrounding the droplet by convection, gas-phase transport processes are presumed to take place within a gas film whose thickness is a function of the droplet relative velocity. This latter assumption, being somewhat inconsistent with the restriction of spherical symmetry, limits the model to either moderate relative velocities of less than 1 m/sec or the forward stagnation region of the droplet. For this initial formulation of the governing equations, the rates of chemical reaction and heat release in the gas phase are assumed to be arbitrary functions of temperature and chemical species.

For the subcritical pressure range of interest, gas-phase heat and mass diffusion rates are of the order of 10^{-1}–1 cm^2/sec while the liquid-phase heat transfer rate is of the order of 10^{-3} cm^2/sec, and the liquid surface area regression rate is approximately 10^{-4}–10^{-3} cm^2/sec. Inasmuch as the gas-phase transfer rates are much faster than all of the liquid-phase transfer rates, gas-phase heat and mass transfer can be represented as quasi-steady processes. The validity of this quasi-steady approximation has been substantiated by the numerical study of Hubbard et al. (*10*). Furthermore, Law and Sirignano (*6*) have demonstrated that effects caused by the liquid surface regression during the droplet heating period are negligible relative to the liquid-phase heat conduction rate.

Based on the above assumptions, the liquid-phase temperature, T_1, is given by the heat conduction equation:

$$\frac{1}{x^2} \frac{\partial}{\partial \varrho} \left(x^2 \frac{\partial T_1}{\partial x} \right) = \frac{a^2}{\alpha_1} \frac{\partial T_1}{\partial t} \tag{1}$$

where the nondimensional radial position is defined for the region $0 \leq x \leq 1$. If spatial variations in the product of mass density and diffusion coefficients (i.e., $\rho\alpha$ and ρD) are small compared with variations in temperature and chemical species, the nondimensional quasi-steady conservation equations for heat and mass transfer within a gas film surrounding the droplet are:

$$\frac{\partial}{\partial x} \left(x^2 \frac{\partial T_g}{\partial x} \right) - \frac{\dot{m}}{4\pi a \rho_g \alpha_g} \frac{\partial T_g}{\partial x} = \frac{a^2 x^2}{k_g} HR \tag{2}$$

where

$$1 \leq x \leq \frac{a + \delta_h}{a}$$

$$\frac{\partial}{\partial x} \left(x^2 \frac{\partial w_f}{\partial x} \right) - \frac{\dot{m}}{4\pi a \rho_g D_f} \frac{\partial w_f}{\partial x} = \frac{a^2 x^2}{\rho_g D_f} R \tag{3}$$

where

$$1 \leq x \leq \frac{a + \delta_f}{a}$$

$$\frac{\partial}{\partial x} \left(x^2 \frac{\partial w_o}{\partial x} \right) - \frac{\dot{m}}{4\pi a \rho_g D_o} \frac{\partial w_o}{\partial x} = \frac{a^2 x^2}{\rho_g D_o} \gamma R \tag{4}$$

where

$$1 \leq x \leq \frac{a + \delta_o}{a}$$

Similar equations could be written for conservation of other reactant or product species. For continuity of temperature, chemical species, heat, and mass at the droplet and film boundaries, the solutions of these governing differential equations are subject to the following boundary conditions:

$$\frac{\partial T_1}{\partial x} (0, t) = 0 \tag{5}$$

$$T_1(1, t) = T_g(1, t) = T_s(t) \tag{6}$$

$$k_1 \frac{\partial T_1}{\partial x}(1, t) = k_g \frac{\partial T_g}{\partial x}(1, t) - \frac{\dot{m}(t) L}{4\pi a(t)} \tag{7}$$

$$T_g \left(\frac{a + \delta_h}{a}, t \right) = T_\infty \tag{8}$$

$$w_f(1, t) = w_{f)s}(t) \tag{9}$$

$$w_f \left(\frac{a + \delta_f}{a}, t \right) = 0 \tag{10}$$

$$\frac{\dot{m}(t)}{4\pi a(t)} w_o(1, t) = \rho_g D_o \frac{\partial w_o}{\partial x}(1, t) \tag{11}$$

$$w_o \left(\frac{a + \delta_o}{a}, t \right) = w_{o)\infty} \tag{12}$$

$$\frac{\dot{m}(t)}{4\pi a(t)} = - \frac{\rho_g D_f}{[1 - w_f(1, t)]} \frac{\partial w_f}{\partial x}(1, t) \tag{13}$$

Initial conditions for the dependent variables are assumed to be arbitrary. By further assuming that the fuel vapor is saturated at the droplet surface, the fuel vapor weight fraction, $w_{f)s}$, and the temperature at the droplet surface, T_s, are related by the Clausius–Clapeyron equation

$$w_{f)s} = \frac{M_f}{M_m} \exp \left[\frac{L M_f}{\mathcal{R}} \left(\frac{1}{T_b} - \frac{1}{T_s} \right) \right] \tag{14}$$

where the molecular weight of the gas mixture, M_m, accounts for the temporal variation of chemical species at the droplet surface.

   With a knowledge of the functional dependence of the rate at which fuel reacts, R, and the physical properties of the reacting chemical species, the droplet combustion system under consideration is completely defined by the above set of simultaneous governing differential equations and boundary conditions.

   Integral Equation Solutions. As a consequence of the quasi-steady approximation for gas-phase transport processes, a rigorous simultaneous solution of the governing differential equations is not necessary. This mathematical simplification permits independent analytical solution of each of the ordinary and partial differential equations for selected boundary conditions. Matching of the remaining boundary condition can be accomplished by an iterative numerical analysis of the solutions to the governing differential equations.

Using Green's function techniques, the heat diffusion equation can be transformed to an implicit integral equation which is amenable to numerical solution. Similar techniques have been used to model both steady-state and transient diffusion processes (*3, 7, 18, 19, 20*). A detailed discussion of the methods for transforming ordinary and partial differential equations to integral equations using Green's function techniques can be found in Ref. *21*. The integral equation for the droplet temperature as a function of position and time obtained from the transformation of Equation 1 for a time-dependent heat flux at the droplet surface is:

$$T_1(x, t) = \frac{1}{4\pi} \int_0^t \frac{\partial T_1}{\partial x_o} (1, t_o)\, G(x, t/1, t_o)\, dt_o \tag{15}$$

$$+ \frac{1}{4\pi} \int_0^1 \frac{a^2}{\alpha_1} x_o{}^2\, T_1(x_o, 0)\, G(x, t/x_o, 0)\, dx_o$$

where the Green's function is:

$$G(x, t/x_0, t_0) = \frac{12\pi\alpha_1}{a^2} + \frac{8\pi\alpha_1}{xx_0a^2} \sum_{n=1}^{\infty} \frac{N_n{}^2 + 1}{N_n{}^2} \sin(N_n x)\sin(N_n x_o)\, e^{-\frac{N_n{}^2\alpha_1(t-t_o)}{a^2}}$$

and the N_n are the positive roots of the characteristic equation:

$$N_n \cot N_n - 1 = 0$$

In Equation 15 the first integral with respect to time represents the effect of the time-dependent heat flux at the droplet surface, and the second integral with respect to position represents the effect of initial conditions. As a result of the form of the integral equation solution in Equation 15, a versatile numerical procedure can be formulated by expressing the transient as a series of successive short-time intervals using a suitable time-averaged value for the heat flux at the droplet surface. In this manner, the droplet temperature profile computed at each value of time becomes the initial condition for the computations pertinent to the subsequent time. The time-averaged liquid temperature gradient required to evaluate the first integral in Equation 15 for each of the successive time intervals is given by Equation 7 for the heat balance condition at the droplet surface. The gas-phase temperature gradient at the droplet surface and the vaporization rate needed to evaluate Equation 7 are obtained from solutions of Equations 2 and 3 for the quasi-steady gas-phase transport processes.

A method for obtaining integral equation solutions to Equations 2, 3, and 4 for the quasi-steady gas-phase temperature and weight fraction profiles is presented in the Appendix. The solution for the quasi-steady gas-phase temperature profile subject to the boundary conditions in Equations 6 and 8 is:

$$T_g(x) = -\int_1^{x_2} \frac{Ha^2 x_o^2}{k_g} e^{\theta_g/x_o} R(x_o, T_g, w_f, w_o) G(x, x_o) dx_o \qquad (16)$$

$$+ T_\infty \left[\frac{e^{-\theta_g/x} - e^{-\theta_g}}{e^{-\theta_g/x_2} - e^{-\theta_g}} \right] + T_s \left[\frac{e^{-\theta_g/x_2} - e^{-\theta_g/x}}{e^{-\theta_g/x_2} - e^{-\theta_g}} \right]$$

where the dimensionless vaporization rate, θ_g, and the dimensionless film position, x_2, are defined, respectively, as:

$$\theta_g = \frac{\dot{m}}{4\pi a \rho_g \alpha_g} \quad \text{and} \quad x_2 = \frac{a + \delta_h}{a}$$

The corresponding solution for the quasi-steady fuel vapor weight fraction in the gas phase subject to the boundary conditions in Equations 9 and 10 is:

$$w_f(x) = -\int_1^{x_2} \frac{a^2 x_o^2}{\rho_g D_f} e^{\theta_f/x_o} R(x_o, T_g, w_f, w_o) G(x, x_o) dx_o \qquad (17)$$

$$+ w_{f)s} \left[\frac{e^{-\theta_f/x_2} - e^{-\theta_f/x}}{e^{-\theta_f/x_2} - e^{-\theta_f}} \right]$$

where

$$\theta_f = \frac{\dot{m}}{4\pi a \rho_g D_f} \quad \text{and} \quad x_2 = \frac{a + \delta_f}{a}$$

Similarly, the quasi-steady oxygen weight fraction profile in the gas phase subject to the boundary conditions in Equations 11 and 12 is:

$$w_o(x) = -\gamma \int_1^{x_2} \frac{a^2 x_o^2}{\rho_g D_o} e^{\theta_o/x_o} R(x_o, T_g, w_f, w_o) G(x, x_o) dx_o \qquad (18)$$

$$+ w_{o)\infty} e^{(\theta_o/x_2 - \theta_o/x)}$$

where

$$\theta_o = \frac{\dot{m}}{4\pi a \rho_g D_o} \quad \text{and} \quad x_2 = \frac{a + \delta_o}{a}$$

The Green's functions, $G(x, x_o)$, needed to evaluate Equations 16, 17, and 18 can be found in the Appendix. The quasi-steady gas-phase temperature and weight fraction profiles for the case of pure vaporization ($R(x, T_g, w_f, w_o) = 0$) are given by the second term of Equations 16, 17, and 18.

Numerical Analysis. To complete the present model, the conditions for ignition, the rate at which fuel is reacted, R, and the film thickness for gas-phase transport, δ, must be known. Because of the difficulties in identifying the numerous intermediate chemical species produced within the gas phase during droplet ignition, an ignition criterion developed by Faeth and Olson (22) which is based on an overall Arrhenius rate law was chosen for this investigation. This ignition criterion was formulated from the point of view that self-ignition occurs when the concentration of a crucial chain-branching intermediate species (unidentified) reaches a critical value in the gas phase. Faeth and Olson showed that the rates of formation of this crucial intermediate species are influenced by the gas-phase history for pressure, temperature, and fuel and oxygen concentrations. During this ignition period, fuel vapor is assumed to diffuse into the surrounding gas phase without appreciable chemical change or heat release; prior to ignition $R = 0$ in Equations 16, 17, and 18. The buildup of this crucial intermediate species relative to the critical concentration of the crucial intermediate at each position in the gas phase is given by:

$$\Psi(x,t) = \frac{X_i(x,t)}{X_{ic}} = \int_0^t K_1 P^{K_2}\phi^{K_3}(x,t) \exp\left[-\frac{E}{\mathcal{R}T_g(x,t)}\right] dt \quad (19)$$

where the rate constants, K_1, K_2, K_3, and E, are empirically determined, and the equivalence ratio is defined as:

$$\phi(x,t) = \gamma\frac{w_f(x,t)}{w_o(x,t)}.$$

The variation of $\Psi(x, t)$ then describes the tendency toward ignition; when it reaches unity, ignition is predicted. An assessment of this simple fuel droplet ignition criterion and appropriate rate constants is contained in Ref. 7.

In keeping with the assumptions of the classical quasi-steady droplet combustion models by Spalding (1) and Godsave (2), the combustion process is assumed for this investigation to take place instantaneously at a flame surface where the fuel and oxygen react stoichiometrically and completely. Then the rate at which fuel is reacted at a flame surface located at $r = b$ can be expressed in terms of the delta function by:

$$R(x) = \frac{\dot{m}}{4\pi b^2}\frac{\delta(x - x_b)}{a} \quad (20)$$

where the delta function is defined as:

$$\delta(x - x_b) = \begin{cases} 0 \text{ for } x \neq x_b \\ \infty \text{ for } x = x_b \end{cases} \qquad \int_0^\infty \delta(x - x_b) F(x)\, dx = F(x_b)$$

and

$$x_b = b/a$$

Also, the flame surface approximation requires that:

$$w_o(x_b, t) = 0 \tag{21}$$

Upon substituting Equation 20 for the reaction rate of fuel into Equations 16, 17, and 18 and performing the integrations, the quasi-steady gas-phase temperature and weight fraction profiles during ignition and combustion are found to be:

$$T_g(x) = T_\infty \left[\frac{e^{-\theta_g/x} - e^{-\theta_g}}{e^{-\theta_g/x_2} - e^{-\theta_g}} \right] + T_s \left[\frac{e^{-\theta_g/x_2} - e^{-\theta_g/x}}{e^{-\theta_g/x_2} - e^{-\theta_g}} \right] \tag{22}$$

$$+ \frac{H}{c_g} \begin{cases} \left[\dfrac{e^{-\theta_g/x} - e^{-\theta_g}}{e^{-\theta_g/x_2} - e^{-\theta_g}} \right] [1 - e^{(\theta_g/x_b - \theta_g/x_2)}] \text{ for } 1 \leq x \leq x_b \\[3ex] \left[\dfrac{e^{-\theta_g/x} - e^{-\theta_g/x_2}}{e^{-\theta_g/x_2} - e^{-\theta_g}} \right] [1 - e^{(\theta_g/x_b - \theta_g)}] \text{ for } x_b \leq x \leq x_2 \end{cases}$$

$$w_f(x) = w_{f)s} \left[\frac{e^{-\theta_f/x_2} - e^{-\theta_f/x}}{e^{-\theta_f/x_2} - e^{-\theta_f}} \right] \tag{23}$$

$$+ \begin{cases} \left[\dfrac{e^{-\theta_f/x} - e^{-\theta_f}}{e^{-\theta_f/x_2} - e^{-\theta_f}} \right] [1 - e^{(\theta_f/x_b - \theta_f/x_2)}] \text{ for } 1 \leq x \leq x_b \\[3ex] \left[\dfrac{e^{-\theta_f/x} - e^{-\theta_f/x_2}}{e^{-\theta_f/x_2} - e^{-\theta_f}} \right] [1 - e^{(\theta_f/x_b - \theta_f)}] \text{ for } x_b \leq x \leq x_2 \end{cases}$$

$$w_o(x) = w_{o)\infty} \, e^{(\theta_o/x_2 - \theta_o/x)} \tag{24}$$

$$- \gamma \begin{cases} e^{-\theta_o/x} [e^{\theta_o/x_b} - e^{\theta_o/x_2}] \text{ for } 1 \leq x \leq x_b \\ [1 - e^{(\theta_o/x_2 - \theta_o/x)}] \text{ for } x_b \leq x \leq x_2 \end{cases}$$

From Equation 22, the temperature gradient in the gas phase at the droplet surface is:

$$\frac{dT_g}{dx} (1) = \frac{\theta_g e^{-\theta_g} \left\{ T_\infty - T_s + \frac{H}{c_g} [1 - e^{(\theta_g/x_b - \theta_g/x_2)}] \right\}}{e^{-\theta_g/x_2} - e^{-\theta_g}} \tag{25}$$

The normalized flame surface position determined from Equations 21 and 24 for the oxygen weight fraction at the flame surface is:

$$x_b = \frac{b}{a} = \frac{\theta_o}{\dfrac{\theta_o}{x_2} - \ln \left[\dfrac{\gamma}{w_{o)\infty} + \gamma} \right]} \tag{26}$$

where

$$x_2 = \frac{\alpha + \delta_o}{a}$$

Then by evaluating Equation 13 for the fuel flux at the droplet surface, using Equation 23 and 26, the nondimensional vaporization rates are found to be:

$$\theta_o = \frac{\dot{m}}{4\pi a \rho_g D_o} = -\left(\frac{x_2}{x_2 - 1} \right) \left\{ \frac{D_f}{D_o} \ln [1 - w_{f)s}] + \ln \left[\frac{\gamma}{w_{o)\infty} + \gamma} \right] \right\} \tag{27}$$

$$\theta_f = \theta_o \frac{D_o}{D_f} \tag{28}$$

$$\theta_g = \theta_o \frac{D_o}{\alpha_g} \tag{29}$$

where

$$x_2 = \frac{a + \delta_o}{a}$$

Equations 22–29 for the gas-phase profiles, temperature gradient, flame position, and vaporization rate depend only on the temperature at the surface of the droplet, T_s. (The Clausius–Clapeyron equation relates $w_{f)s}$ to T_s.) An iterative numerical procedure for satisfying the continuity of T_s at the liquid/gas interface is described in the Appendix.

Estimates of the film thicknesses, δ_j, needed to determine the gas-phase temperature and weight fraction profiles are based on the empirical Nusselt number correlations developed by Ranz and Marshall (*23*) for simultaneous heat and mass transfer during droplet vaporization. A detailed derivation of the equations for the gas-phase film thicknesses is available in Ref. 7. This derivation consists of equating the heat and mass fluxes at the droplet surface based on the profiles given by Equations 22, 23, and 24 to the corresponding fluxes expressed in terms of the Nusselt numbers for free or forced convection. By properly accounting for the effects of droplet vaporization, unidirectional mass transfer, and chemical reaction, the expression obtained for the gas-phase film thicknesses is:

$$\frac{\delta_j}{a} = \frac{2}{\mathrm{Nu} - 2} \tag{30}$$

The empirical correlations developed by Ranz and Marshall (23) for the Nusselt numbers for simultaneous heat and mass transfer during droplet vaporization are as follows:
For forced convection of heat and mass:

$$\mathrm{Nu_h} = 2 + 0.6\ \mathrm{Re}^{1/2}\ \mathrm{Pr}^{1/3} \tag{31a}$$

$$\mathrm{Nu_m} = 2 + 0.6\ \mathrm{Re}^{1/2}\ \mathrm{Sc}^{1/3} \tag{31b}$$

For free convection of heat and mass:

$$\mathrm{Nu_h} = 2 + 0.6\ \mathrm{Gr}^{1/4}\ \mathrm{Pr}^{1/3} \tag{32a}$$

$$\mathrm{Nu_m} = 2 + 0.6\ \mathrm{Gr}^{1/4}\ \mathrm{Sc}^{1/3} \tag{32b}$$

where

$$\mathrm{Nu_h} = 2ah_c/k_g, \ \mathrm{Nu_m} = 2ak_c/\rho_g D_j$$

$$\mathrm{Re} = 2a\rho_g V_r/\mu_g, \ \mathrm{Pr} = \mu_g c_g/k_g$$

$$\mathrm{Gr} = (2a)^3 \rho_g^2 g\ \Delta T/\mu g, \ \mathrm{Sc} = \mu_g/\rho_g D_j$$

For the present investigation the characteristic temperature difference for free convection, ΔT, is taken as $|T_\infty - T_s|$, and the gas-phase properties are evaluated at the mean temperature $\overline{T_g} = 1/2(T_\infty - T_s)$. The thermal coefficient of volumetric expansion of the gas, β, which appears in the Grashof number is taken as $1/\overline{T_g}$. Equations 30, 31, and 32 state that the gas film thickness which surrounds the droplet is infinite when the droplet is motionless relative to the gas stream and when gravity is absent. As the relative velocity increases, the film thickness becomes smaller.

Results and Discussion

The present theoretical model is used here to analyze the entire ignition and combustion process for a single droplet and to examine the effects of nonsteady droplet heating and droplet motion relative to the hot gas environment. The computations pertain to a 300 μ furfuryl alcohol droplet with an initial droplet temperature of 295 K, and the conditions for the hot gas environment are taken to be $T_\infty = 1400$ K and $w_{o)\infty} = 0.1355$ g O_2/g air. The rate constants used for the ignition criterion are those determined in Ref. 7. The effect of droplet relative motion on ignition and combustion is illustrated in Figures 2–6 as a function of the gas-phase heat and mass transfer mode; namely, pure diffusion, free convection, and forced convection with the droplet relative velocity taken as $|V_d - V_\infty| = 25$ cm/sec.

The ignition criterion profiles at the time of ignition when $\Psi(x, t)$ first reaches unity are shown in Figure 2 for the three different modes of gas-phase heat and mass transfer. Both the time and position in the

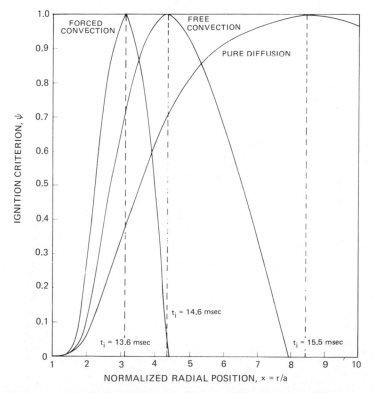

Figure 2. Ignition criterion profiles as a function of gas-phase transfer mode. For a 300-μ furfuryl alcohol droplet in air at 1400 K and 0.1355 g O_2/g air.

gas phase where self-ignition takes place decrease with increasing convection strength. While ignition delay times with free or forced convection are only 6–12% less than the ignition time for pure diffusion, the positions of ignition with free or forced convection are two to three times closer to the droplet surface than for pure diffusion.

Figure 3 shows the temporal variations of the normalized droplet surface area and the rate of droplet vaporization for the three modes of gas-phase heat and mass transfer. Surprisingly there is very little influence

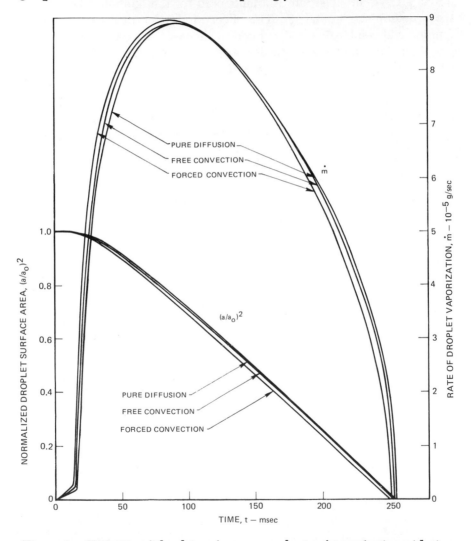

Figure 3.   Variation of droplet surface area and rate of vaporization with time as a function of gas-phase transfer mode. Ignition and combustion of a 300-μ furfuryl alcohol droplet in air at 1400 K and 0.1355 g O_2/g air.

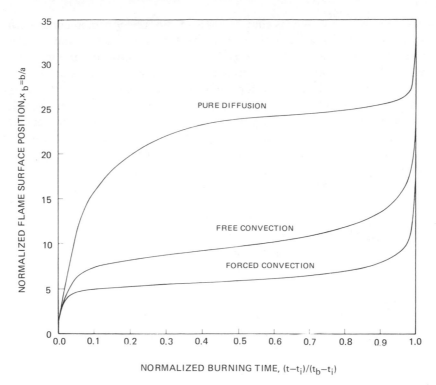

Figure 4.   Variation of flame surface position with time as a function of gas-phase transfer mode.   Combustion of a 300-μ furfuryl alcohol droplet in air at 1400 K and 0.1355 g O_2/g air.

of gas-phase convection on the rate of droplet vaporization during both the ignition and combustion periods. While at early times the vaporization rate is higher for the convection modes as compared with pure diffusion, the trend reverses at later times. This effect can be explained by noting that at later times for pure diffusion a larger droplet size produces a higher rate of vaporization as compared with the convection modes since the vaporization rate is proportional to the instantaneous droplet radius (*see* Equation 27). The nonsteady heating of the droplet is exhibited in Figure 3 by the curvature in the graph for the normalized droplet surface area during the initial phases of ignition and combustion. Figure 3 also shows that very little droplet vaporization occurs prior to ignition.

The temporal variation in the normalized flame surface position is shown in Figure 4 to be affected significantly by the mode of gas-phase heat and mass transfer. Although this appears to be inconsistent with the trend illustrated in Figure 2 for the rate of droplet vaporization, the paradox will be explained later by the influence of convection on the gas-phase heat flux at the droplet surface. As a result of the finite gas-phase film thickness which is characteristic of the free or forced convection modes, the normalized flame position is substantially closer to the droplet surface with convection than it is for the pure diffusion mode. The functional dependence of the normalized flame surface position, x_b, on the gas-phase film thickness, δ, given by Equation 26 shows that as the normalized gas-phase film thickness, x_2, decreases (as with convection), the normalized flame position, x_b, also decreases. The substantial increase in the normalized flame surface position predicted in Figure 4 during combustion is qualitatively consistent with the observations of Nuruzzaman et al. (11) for the combustion of mono-sized droplet streams.

A comparison of the relative position in the gas phase where ignition is predicted by the ignition criterion, Ψ, (Figure 2) with the normalized flame position, x_b, (Figure 4) at the start of combustion for corresponding modes of gas-phase heat and mass transfer suggests that the flame surface rapidly collapses toward the droplet surface after ignition, and then the

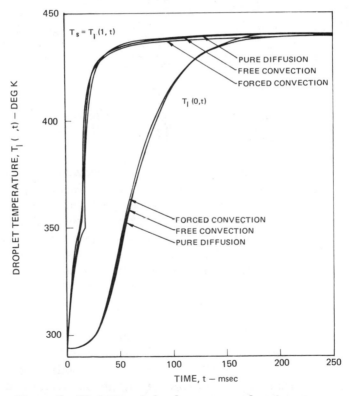

Figure 5.   Variation of droplet center and surface temperatures with time as a function of gas-phase transfer mode. Ignition and combustion of a 300-μ furfuryl alcohol droplet in and at 1400 K and 0.1355 g O₂/g air.

flame surface expands away from the droplet during combustion. Although this discontinuity may be attributed partially to the fact that completely different methods are used to evaluate the ignition and combustion periods, the rather large quantitative differences between the flame positions predicted immediately before and after ignition lends considerable credence to the postulate of a collapsing and expanding flame surface. However, a conclusive examination of this postulate regarding the flame surface position immediately following ignition requires a much more complex representation of the combustion chemistry than the ignition criterion and flame surface approximation used in this investigation.

The temporal variation of the temperatures at the center and surface of the droplet are shown in Figure 5 to be nearly independent of the mode of gas-phase heat and mass transfer. While the surface temperature increases rapidly during the initial period of ignition and combustion, the center temperature increases at a much slower rate. This trend of non-uniform heating is characteristic of liquids, such as furfuryl alcohol, having a high heat capacitance. The droplet temperature never reaches the boiling temperature (444 K) throughout the entire combustion process. Figure 5 also shows that much of the temperature rise at the droplet surface takes place during the preignition period.

The gas-phase temperature profiles at selected times during the ignition and combustion processes are shown in Figure 6 for the three modes of gas-phase heat and mass transfer. The most notable effects of enhanced gas-phase convection, shown for each value of time, are a compressing of the temperature profiles closer to the droplet surface and a reduction in the flame surface temperature. Figure 6 also shows that convection steepens the temperature gradient at the gas film boundary such that a greater percentage of the heat produced by combustion is transferred to the environment with either convection mode than by pure diffusion alone. Although the flame position is closer to the droplet surface with free or forced convection, respectively, than for pure diffusion, the rate of heat transfer to the droplet surface at any time is essentially the same for the three modes of gas-phase transport considered because the flame surface temperature is reduced with enhanced gas-phase transfer. Hence the droplet vaporization rate should be nearly independent of the mode of gas-phase transfer as shown in Figure 3. This effect of convection on the vaporization rate can be understood more clearly by examining Equation 25 for the temperature gradient in the gas phase at the droplet surface. Compared with the normalized flame position, x_b, the normalized boundary position, x_2, is infinite for pure diffusion and finite for free or forced convection. Equation 25 shows that as x_2 approaches x_b, as happens with increased convection, the percentage of the heat

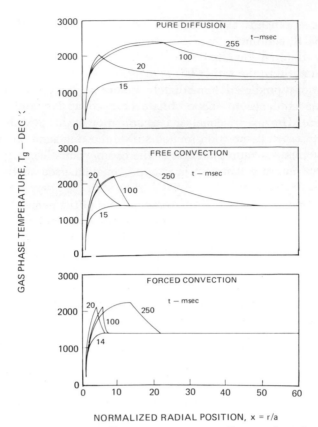

Figure 6.   Gas-phase temperature profiles at selected times as a function of gas-phase transfer mode. Ignition and combustion of a 300-μ furfuryl alcohol droplet in air at 1400 K and 0.1355 g O₂/g air.

produced by combustion which is transferred to the droplet surface decreases. In fact, when $x_2 = x_b$, Equation 25 indicates that none of the combustion heat is transferred to the droplet. This implies that the droplet flame will extinguish if $x_2 = x_b$ since all of the combustion heat will be transferred to the environment.

A comparison of the present theoretical model with the experimental observations of Okajima and Kumagai (8) is shown in Figure 7 for the combustion of a 1300-μ ethyl alcohol droplet under the influence of weak forced convection as a result of relative motion with respect to the gas phase in a gravity-free environment not disturbed by free convection. For these experiments Okajima and Kumagai (8) used a freely falling chamber within which a free droplet was combusted while the droplet moved at a relative velocity of 1.29 cm/sec in a standard air atmosphere with an initial temperature of 300 K. Since the droplet was spark-ignited for these experiments, the initial droplet temperature for the analysis is taken as 300 K, and the ignition process is not considered. The combustion process is assumed for this analysis to take place instantaneously at a flame surface. Figure 7 shows comparisons of the predicted and observed temporal variations of the normalized droplet surface area and the normalized flame dimensions. The system for measuring the flame dimensions is shown in Figure 7 where the measured dimensions for the observed elliptically shaped flame are H_1, H_2, and W, and the predicted dimension for the spherically symmetric flame is b. The temporal variation of droplet surface area indicates that the effects of transient droplet heating exist for approximately the first 10% of the droplet lifetime; afterward the burning process assumes the quasi-steady behavior.

Close agreement of the observed and predicted variations for the normalized droplet surface area are shown in Figure 7 for both pure diffusion and forced-convection modes of gas-phase heat and mass transfer. Whereas the consistency of normalized surface area predictions for both

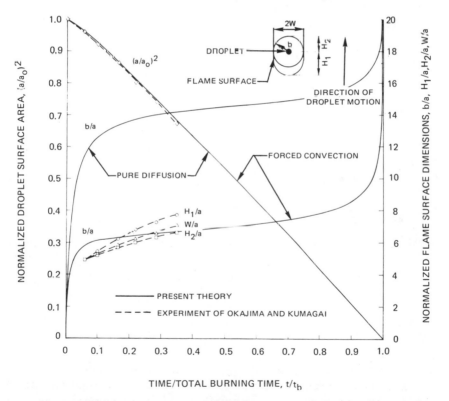

Figure 7. Comparison of predicted and observed variation of droplet surface area and flame surface position. Combustion of a 1300-μ ethyl alcohol droplet in a standard air atmosphere at 300 K with $|V_d - V_\infty| = 1.29$ cm/sec.

modes of gas-phase transfer is not surprising in view of the predicted results shown in the previous figures, the good agreement of these predic- itons with the observed surface area variation partly results from the selection of physical and chemical constants to simulate effects of variable properties by matching the total burning time. With regard to flame dimensions, the prediction based on the forced convection mode of gas- phase transfer agrees very well with the observed flame dimensions, particularly for the forward stagnation region of the droplet. However, the prediction for flame surface based on the pure diffusion mode of gas-phase transfer deviates considerably from the observed dimensions just as has been reported for previous theoretical models using a gas-phase boundary of infinite extent.

In conclusion, the results reported above indicate that convective transfer of heat and mass in the gas phase surrounding a burning droplet can significantly affect the flame dimensions. Even very small levels of relative motion, of the order of 1 cm/sec such as for the situation illus- trated in Figure 7, can alter the flame dimensions by a significant factor. However, droplet burning rate is shown to be only slightly affected by the mode of gas-phase heat and mass transfer.

Appendix

Integral Equations for Representing Gas-Phase Temperature and Weight Fraction Profiles. The details are presented in this section for transforming the quasi-steady differential equations for conservation of gas-phase temperature and weight fractions to integral equations using Green's function techniques. The differential equations considered here differ from the equations for the classical quasi-steady droplet combus- tion models by Spalding (*1*) and Godsave (*2*) in that the gas-phase region is of finite thickness, δ, and the reaction rate of the fuel, *R*, is considered to be an arbitrary function of temperature and chemical species concentrations.

The differential equations to be solved are of the form:

$$\frac{d}{dx}\left(x^2 \frac{dF}{dx}\right) - \theta \frac{dF}{dx} = \xi \qquad (A\text{-}1)$$

where the dependent variable F represents either the gas-phase tempera- ture or a chemical species weight fraction, θ is the dimensionless rate of vaporization, ξ is the reaction rate function, and the dimensionless position is defined for the region $x_1 \leq x \leq x_2$.

Introducing the integrating factor:

$$S = \exp\left(-\int_x^\infty \frac{\theta}{\lambda^2}\, d\lambda\right)$$

and assuming that the dimensionless rate of vaporization, θ, is spatially uniform compared with F, such that $S = e^{\theta/x}$, Equation A-1 can be rewritten as:

$$\frac{d}{dx}\left(e^{\theta/x}\, x^2\, \frac{dF}{dx}\right) = e^{\theta/x}\, \xi \tag{A-2}$$

An analogous Green's function is chosen to satisfy the differential:

$$\frac{d}{dx}\left(e^{\theta/x}\, x^2\, \frac{dG}{dx}\right) = -\,\delta(x - x_o) \tag{A-3}$$

where $x_1 \le x \le x_2$, and the delta function has the properties:

$$\delta(x - x_o) = \begin{cases} 0 \text{ for } x \ne x_o \\ \infty \text{ for } x = x_o \end{cases} \quad \text{and} \quad \int_{x_1}^{x_2} F(x_o)\,\delta(x - x_o)\, dx_o = F(x)$$
$$\text{for } x_1 \le x \le x_2$$

The Green's function $G = G(x, x_o)$ describes the scalar field at a position x as a result of a unit impulsive source at some other position x_o. Hence, G is a function of an observer system x and a source system x_o. If x_o is chosen within the same interval as for x, Equations A-2 and A-3 can be rewritten in terms of the source coordinate system simply by replacing x by x_o. Then multiplying Equation A-2 by G and Equation A-3 by F, subtracting the products, and integrating over the source coordinate, x_o, from x_1 to x_2 gives:

$$\int_{x_1}^{x_2} \frac{d}{dx_o}\left(F\, e^{\theta/x_o}\, x_o^2\, \frac{dG}{dx_o} - G\, e^{\theta/x_o}\, x_o^2\, \frac{dF}{dx_o}\right) dx_o \tag{A-4}$$

$$+ \int_{x_1}^{x_2} F(x_o)\,\delta(x - x_o)\, dx_o + \int_{x_1}^{x_2} e^{\theta/x_o}\, \xi(x_o, F)\, G(x, x_o)\, dx_o = 0$$

Using integration by parts to evaluate the first integral, recalling that θ is taken as spatially uniform, and using the properties of the delta function to evaluate the second integral, the implicit integral equation for F is:

$$F(x) = -\int_{x_1}^{x_2} e^{\theta/x_o}\, \xi(x_o, F)\, G(x, x_o)\, dx_o \tag{A-5}$$

$$-\left[F e^{\theta/x_o}\, x_o^2\, \frac{dG}{dx_o} - G\, e^{\theta/x_o}\, x_o^2\, \frac{dF}{dx_o}\right]_{x_o = x_1}^{x_o = x_2}$$

In Equation A-5 the first term represents the effects of the arbitrary reaction rate function, $\xi(x_0, F)$, and the second term represents the effects of boundary conditions.

The boundary conditions to be satisfied by the Green's function are chosen as the homogeneous conditions corresponding to the conditions for F. As an example, the analogous boundary conditions selected for gas-phase temperature and fuel vapor weight fraction in the present model are:

$$F(x_1) = F_1 \tag{A-6a}$$

$$F(x_2) = F_2 \tag{A-6b}$$

$$G(x,x_1) = 0 \tag{A-6c}$$

$$G(x,x_2) = 0 \tag{A-6d}$$

Upon substituting these boundary conditions in Equation A-6 into Equation A-5, the integral equation solution is:

$$F(x) = -\int_{x_1}^{x_2} e^{\theta/x_0}\, \xi(x_0,F)\, G(x,x_0)\; dx_0 \tag{A-7}$$

$$+ F_1\, e^{\theta/x_1}\, x_1^2\, \frac{dG}{dx_0}(x,x_1) - F_2\, e^{\theta/x_2}\, x_2^2\, \frac{dG}{dx_0}(x,x_1)$$

The solution of Equation A-3 for the Green's function, subject to the boundary conditions in Equations A-6c and A-6d, is:

$$G(x,x_0) = \begin{cases} \dfrac{(e^{-\theta/x} - e^{-\theta/x_1})\,(e^{-\theta/x_2} - e^{-\theta/x_0})}{\theta\,(e^{-/\theta x_2} - e^{-\theta/x_1})} & \text{for } x \le x_0 \\[2em] \dfrac{(e^{-\theta/x} - e^{-\theta/x_2})\,(e^{-\theta/x_1} - e^{-\theta/x_0})}{\theta\,(e^{-\theta/x_2} - e^{-\theta/x_1})} & \text{for } x \ge x_0 \end{cases} \tag{A-8}$$

and

$$\frac{dG}{dx_0}(x,x_0) = \begin{cases} \dfrac{-\,e^{-\theta/x_0}\,(e^{-\theta/x_1} - e^{-\theta/x_1})}{x_0^2\,(e^{-\theta/x_2} - e^{-\theta/x_1})} & \text{for } x \le x_0 \\[2em] \dfrac{-\,e^{-\theta/x_0}\,(e^{-\theta/x} - e^{-\theta/x_2})}{x_0^2\,(e^{-\theta/x_2} - e^{-\theta/x_1})} & \text{for } x \ge x_0 \end{cases} \tag{A-9}$$

The analogous boundary conditions selected for oxygen weight fraction in the present model are:

$$\frac{dF}{dx_o}(x_1) = \theta F(x_1) \tag{A-10a}$$

$$F(x_2) = F_2 \tag{A-10b}$$

$$\frac{dG}{dx_o}(x,x_1) = \theta G(x,x_1) \tag{A-10c}$$

$$G(x, x_2) = 0 \tag{A-10d}$$

The integral equation solution resulting from the substitution of boundary conditions Equation A-10 into Equation A-5 is:

$$F(x) = \int_{x_1}^{x_2} e^{\theta/x_o}\, \xi(x_o, F)\, G(x, x_o)\, dx_o - F_2\, e^{\theta/x^2} x_2{}^2\, \frac{dG}{dx_o}(x, x_2) \tag{A-11}$$

The Green's function obtained by solving Equation A-3 subject to the boundary conditions in Equations A-10c and A-10d is:

$$G(x, x_o) = \begin{cases} \dfrac{e^{-\theta/x}}{\theta}\left[1 - e^{(\theta/x_2 - \theta/x_o)}\right] \text{ for } x \leq x_o \\[3mm] \dfrac{e^{-\theta/x_o}}{\theta}\left[1 - e^{(\theta/x_2 - \theta/x)}\right] \text{ for } x \geq x_o \end{cases} \tag{A-12}$$

and

$$\frac{dG}{dx_o}(x, x_o) = \begin{cases} -\dfrac{e^{(\theta/x_2 - \theta/x - \theta/x_o)}}{x_o{}^2} \text{ for } x \leq x_o \\[3mm] \dfrac{e^{-\theta/x_o}}{x_o{}^2}\left[1 - e^{(\theta/x_2 - \theta/x)}\right] \text{ for } x \geq x_o \end{cases} \tag{A-13}$$

Numerical Procedure. This section describes an iterative numerical procedure for evaluating the mathematical model presented in this chapter for the nonsteady ignition and combustion of a fuel droplet. The key step of this procedure is to match or couple the analytical equations for heat and mass transport at the liquid/gas interface by establishing the droplet surface temperature for a series of successive short time intervals. The numerical procedure for each time consists of the following steps:

1. Assume the droplet surface temperature, T_s.

2. Compute the fuel vapor weight fraction at the droplet surface, $w_{f)s}$, using Equation 14.

3. Compute the gas-phase film thicknesses for heat and mass transfer, δ_j, using Equations 30, 31, and 32.

4. Compute the nondimensional rates of vaporization, θ_j, using Equations 27, 28, and 29.

5. If ignition has been established, compute the nondimensional flame surface position, x_b, using Equation 26.

6. Compute the gas-phase temperature gradient at the droplet surface, dT_g/dx, using Equation 25.

7. Compute the liquid-phase temperature gradient at the droplet surface, dT_l/dx, using Equation 7.

8. Compute the droplet surface temperature, $T_s = T_1(1, t)$, using Equation 15 with the numerical average of the liquid-phase temperature gradient at the droplet surface from step 7 for this and the previous time.

9. Compare the assumed and computed values for T_s from steps 1 and 8, repeating the above procedure until a satisfactory accuracy is obtained.

10. After establishing the droplet surface temperature at this time, profiles for droplet temperature, gas-phase temperature, fuel weight fraction, and oxygen weight fraction are computed using Equations 15, 22, 23, and 24, respectively.

11. If self-ignition has not yet been established, the profile for the ignition criterion, Ψ, is computed using Equation 19.

12. The droplet size at this time is determined from a finite-difference solution of a droplet mass balance equation using the rate of vaporization computed in step 4.

Since predictions of droplet combustion theories depend to a large extent on the values of the physical properties, the computations for the present investigation use physical properties based on the reference temrature methods suggested by Hubbard (*10*) and Law and Law (*24*).

Nomenclature

a	Radius of spherical fuel droplet, cm
b	Radius of flame surface position, cm
c	Specific heat at constant pressure, cal/g-K
D	Mass diffusion coefficient in gas phase, cm²/sec
E	Activation energy for ignition criterion, cal/g-mol
F	General function
G	Green's function
Gr	Grashof number for heat or mass convection
h_c	Heat convection coefficient, cal/cm²-sec-K
H	Heat of reaction per unit mass of fuel (negative for exothermic reaction), cal/g
k	Thermal conductivity, cal/cm-sec-K
k_c	Mass convection coefficient, cm/sec
K_1	Rate constant in ignition criterion, sec^{-1}

K_2 Constant in ignition criterion denoting order of pressure dependence

K_3 Constant in ignition criterion denoting order of equivalence ratio dependence

L Latent heat of vaporization of liquid fuel, cal/g

$\dot{m}$ Rate of fuel vaporization, g/sec

M Molecular weight

n Integer values, 1, 2, 3, - - - etc.

N_n Positive roots of characteristic equation

Nu   Nusselt number for heat or mass convection

P Pressure, atm

Pr   Prandtl number

r Radial position measured from center of fuel droplet, cm

R Rate of reaction of fuel, g/cm³-sec

$\mathcal{R}$ Perfect gas law constant, equals 1.987 cal/g-mole-K

Re   Reynolds number

S Integrating factor

Sc   Schmidt number

t Time, sec

T Temperature, K

V Velocity, cm/sec

w Weight fraction

x Normalized radial position, equals r/a

X Mole fraction of crucial intermediate

Greek Symbols

α Thermal diffusivity, equals $k/\rho c$, cm²/sec

β Thermal coefficient of volumetric expansion of the gas phase

γ Mass of oxygen per mass of fuel for stoichiometric reaction

δ Gas-phase film thickness for heat or mass convection, cm, or delta function

θ Dimensionless rate of vaporization defined in text

λ Dummy variable

μ Dynamic viscosity, poises

ξ General reaction rate function

ρ Mass density, g/cm³

ϕ Equivalence ratio

Ψ Ignition criterion, defined by Eq. (19)

Subscripts

1   Refers to dimensionless position of inner boundary of gas phase

2   Refers to dimensionless position of outer boundary of gas phase

b   Refers to boiling point of liquid at 1 atm, or to dimensionless flame position

c   Refers to heat or mass convection or to critical value of intermediate
         species
d   Refers to fuel droplet
f   Refers to fuel vapor
g   Refers to gas phase
h   Refers to gas-phase heat transfer
i   Refers to a crucial intermediate or ignition time
j   Refers to a general chemical species or temperature
l   Refers to liquid fuel
m   Refers to gas-phase mass transfer or to gas mixture
n   Refers to integer value, 1, 2, 3, - - - etc.
o   Refers to oxygen, or the source coordinate system (x, x_o) of the
         Green's function
s   Refers to droplet surface
∞   Refers to ambient conditions in the gas phase

Acknowledgments

The author wishes to acknowledge the key contribution of C. J. Hewett who prepared the computer program and was instrumental in developing the numerical procedure for solving the governing integral equations. The many enlightening and helpful discussions with A. S. Kesten are gratefully appreciated.

Literature Cited

1. Spalding, D. B., "Combustion of Fuel Particles," *Fuel* (1951) **30**, 121–130.
2. Godsave, G. A. E., "Studies of the Combustion of Drops in a Fuel Spray," Fourth Symposium on Combustion, pp. 818–830, Williams and Wilkins, Baltimore, 1953.
3. Kesten, A. S., Sangiovanni, J. J., "Temperature Field Development Around Burning Fuel Droplets and Nitric Oxide Production," *Proc. Int. Heat Transfer Conf. 5th* (September 1974).
4. Crespo, A., Linan, A., "Unsteady Effects in Droplet Evaporation and Combustion," *Combust. Sci. Technol.* (1975) **2**, 9.
5. Law, . K., "Unsteady Droplet Combustion with Droplet Heating," *Combust. Flame* (1976) **26**, 17–22.
6. Law, C. K., Sirignano, W. A., "Unsteady Droplet Combustion with Droplet Heating. II: Conduction Limit." *Combust. Sci. Technol.*, in press.
7. Sangiovanni, J. J., Kesten, A. S., "A Theoretical and Experimental Investigation of the Ignition of Fuel Droplets," *Combust. Sci. Technol.*, in press.
8. Okajima, S., Kumagai, S., "Further Investigations of Combustion of Free Droplets in a Freely Falling Chamber Including Moving Droplets," *Symp. (Int.) Combust., 15th (Proc.)* (1975) 401.
9. Kotake, S., Okazaki, T., "Evaporation and Combustion of a Fuel Droplet," *Int. J. Heat Mass Transfer,* (1969) **12**, 595–609.
10. Hubbard, G. L., Denny, V. E., Mills, A. F., "Droplet Evaporation: Effects of Transients and Variable Properties," *Int. J. Heat Mass Transfer* (1975) **18**, 1003–1008.

11. Nuruzzaman, A. S. M., Hedley, A. B., Beer, J. M., "Combustion of Mono-sized Droplet Streams in Stationary Self-Supporting Flames," *Symp. (Int.) Combust., 13th (Proc.)* (1971).
12. Chigier, N. A., McCreath, C. G., Makepeace, R. W., "Dynamics of Droplets in Burning and Isothermal Kerosene Sprays," *Combust. Flame* (1974) **23**, 11–16.
13. Onuma, Y., Ogasawara, M., "Studies of the Structure of a Spray Combustion Flame," *Symp. (Int.) Combust., 15th (Proc.)* (1975).
14. Brzustowski, T. A., Natarajan, R., "Combustion of Aniline Droplets at High Pressures," *Can. J. Chem. Eng.* (1966) **44**, 194.
15. Fendell, F. E., Sprankle, M. L., Dodson, D. S., "Thin-Flame Theory for a Fuel Droplet in Slow Viscous Flow," *J. Fluid Mechan.* (1966) **26**, 267–280.
16. Isoda, H., Kumagai, S., "New Aspects of Droplet Combustion," *Symp. (Int.) Combust., 7th (Proc.)* (1967) 726.
17. El-Wakil, M. M., Priem, R. J., Brikowski, H. J., Myers, P. S., Uyehara, O. A., "Experimental and Calculated Temperature and Mass Histories of Vaporizing Fuel Drops," *NACA TN-3490, 1956.*
18. Sangiovanni, J. J., Kesten, A. S., "Analysis of Gas Pressure Buildup within a Porous Catalyst Particle which is Wet by a Liquid Reactant," *Chem. Eng. Sci.* (1971) **26**, 533–547.
19. Kesten, A. S., "Analysis of NO Formation in Single Droplet Combustion," *Combust. Sci. Technol.* (1972) **6**, 115–123.
20. Sangiovanni, J. J., "An Integral Equation Method for Modeling the Transient Behavior of Catalytic Reactions within Porous Particles," Ph.D. Thesis, Rensselaer Polytechnic Institute, Troy, N.Y., 1974.
21. Morse, P. M., Feshback, H., "Methods of Theoretical Physics, Part I," McGraw-Hill, New York, 1953.
22. Faeth, G. M., Olson, D. R., "The Ignition of Hydrocarbon Fuel Droplets in Air," *SAE Trans.* (1968) **73**, 1793.
23. Ranz, W. E., Marshall, Jr., W. R., "Evaporation from Drops, Parts I and II," *Chem. Eng. Prog.* (1952) **48**, 141–146, 173–180.
24. Law, C. K., Law, H. K., "Quasi-Steady Diffusion Flame Theory with Variable Specific Heats and Transport Coefficients," *Combust. Sci. Technol.* (1976) **12**, 207–217.

RECEIVED December 29, 1976.

3

Drop Interaction in a Spray

ALLEN L. WILLIAMS—R. R. 1, Box 90, Edgewood, IL 62426

JOHN C. CARSTENS—Physics Department, Graduate Center for Cloud Physics Research, University of Missouri—Rolla, Rolla, MO 65401

JOSEPH T. ZUNG—Chemistry Department, University of Missouri—Rolla, Rolla, MO 65401

An analytic formalism describing the direct interaction between growing or evaporating drops in the continuum regime is developed. This formalism is modified for the case in which one of the drops is replaced by an inert sphere. Emphasis is placed on the exposition of appropriate boundary conditions and general solution. There is less than a 10% reduction in the growth/evaporation rate if drop separations exceed about 10 times the average radius. When gravity is the predominant force, relative fall velocities reduce the overall effect of the interaction so that only drops of equal radii suffer prolonged interaction. Thus, the brevity of this interaction between drops of different size tends to minimize any size distribution spreading.

The effects of drop interaction in a spray or cloud are conventionally taken into account by evaluating the influence of the drop assemblage on the overall temperature and vapor concentration or vapor pressure fields as if these were uniform and the drops were otherwise isolated. It follows that in this treatment, the drops interact solely by means of their mutual effect on the assumed uniform vapor and temperature fields. As drop concentration increases, one expects a direct interaction involving the local fields of closely spaced drops; such spatial interaction manifests itself first in terms of drop pairs, then triplets, and so on.

In this chapter we develop formulas describing local vapor concentration and temperature fields for two juxtaposed drops in a gas–vapor mixture of infinite extent. Instantaneous growth or evaporation rates can be calculated straightforwardly from knowledge of the vapor concentration fields in the vicinity of the drop. Our primary and fundamental assumptions are that continuum transport theory is a valid approximation, that the vapor is dilute with respect to the noncondensible carrier gas, that the calculated profiles are adequately represented by steady-state solutions to the relevant transport equations, and that, assuming a gravitational field, fall velocity can be decoupled from the transport processes herein contemplated. Further discussion as well as identification of secondary assumptions appears below.

Basic Equations

The gas mixture under consideration is composed of an indifferent carrier gas and dilute condensate. The total molar concentration of the mixture is c, the binary diffusion coefficient is D, and the thermal conductivity is K.

As we are considering an evaporation/growth process that is predominantly diffusion controlled, we adopt Fick's and Fourier's laws in the following form (1):

$$\vec{N} = -cD\nabla x + x(\vec{N} + \vec{N}_{\text{g}}) \tag{1}$$

$$\vec{\epsilon} = -K\nabla T + \vec{N}\tilde{C}_{\text{p}}\,(T - T_\infty) \tag{2}$$

in which $\vec{N}$ is the molar vapor flux and $\vec{N}_{\text{g}}$ that of the carrier gas, x is the vapor mole fraction, ϵ is the heat flux, T is the temperature (T_∞ that at $r = \infty$, i.e., ambient), and $\tilde{C}_{\text{p}}$ is the molar specific heat at constant pressure.

Our first approximation neglects both convective terms (far right-hand terms). Two major contributions to the convective flux can be identified: the convective field resulting from drop fall and that resulting from the production (evaporation) or adsorption (condensation) of vapor at the surface of the drop. The former we neglect since drop sizes of interest have less than a 30-μ radius so that the effects of fall are likely to contribute only a few percent to the total flux (2). The latter contribution affects only $\vec{N}$ so that with drop fall negelected ($\vec{N}_{\text{g}} = 0$), i.e.:

$$\vec{N} = -[cD/(1-x)]\,\nabla x = cD\nabla \ln(1-x) \tag{3}$$

Now $$\ln(1-x) = x - (x^2/2) + \ldots,$$

and we regard the condensate dilute enough that:

$$\vec{N} \cong -cD\nabla x \tag{4}$$

is satisfactory for our purposes. The vapor flux is likewise neglected in Equation 2. (*See* for example Bird et al. (*1*)). It follows that:

$$\tilde{\epsilon} \cong -K\nabla T \tag{5}$$

Our next approximation considers the transport process as quasi-steady state. This implies that the characteristic diffusion and conduction times, l^2/D and $l^2 c c_p/K$ (in which l is a characteristic length), are much smaller than the characteristic time appropriate for relative fall. A very conservative characteristic time would be that required for a drop to fall a distance of one diameter, i.e., by using Stoke's law:

$$\tau = (9\eta_g/\rho_1 g) \ (1/a) \tag{6}$$

in which a is the drop radius, ρ_1 is the liquid density (assumed to be much greater than that of the gas), g is the acceleration of gravity, and η_g is the gas viscosity. Actually, we are more concerned with drops that are nearly the same size so that we can entertain rather larger values of τ given by:

$$\tau' = (9/2) \ (\eta_g/\rho_1 g) \ (1/\Delta a) \tag{7}$$

in which Δa is the difference in radii. With the above approximations, the equations to be solved are:

$$\nabla^2 x = 0 \tag{8}$$

$$\nabla^2 T = 0 \tag{9}$$

Solutions: Boundary Conditions

The two-sphere geometry suggests use of the bispherical coordinate grid. We adopt the notation of Morse and Feshbach (*3*) to whom we refer for details. The two independent spatial variables describing the axisymmetric geometry are μ and η. In particular μ_1 describes the surface of drop 1 and $-\mu_2$ that of drop 2. (The minus sign associated with μ_2 is carried along explicitly so that μ_2 is a positive number.)

The exterior solutions of Equations 8 and 9 are:

$$T_e = T_\infty + f^{1/2} \sum_{n=0}^{\infty} \{A_n \exp\left[-(n+1/2)\,\mu\right]$$

$$+ B_n \exp\left[(n+1/2)\mu\right]\} P_n(w) \tag{10}$$

$$x = x_\infty + f^{1/2} \sum_{n=0}^{\infty} \{ C_n \exp[-(n+1/2)\,\mu]$$
$$+ D_n \exp[(n+1/2)\,\mu] \} P_n(w) \tag{11}$$

in which $-\mu_2 \le \mu \le \mu_1$. The interior solutions are:

$$T_{i1} = f^{1/2} \sum_{n=0}^{\infty} F_{1n} \exp[-(n+1/2)\,\mu] P_n(w) \tag{12}$$

$$T_{i2} = f^{1/2} \sum_{n=0}^{\infty} F_{2n} \exp[-(n+1/2)\,\mu] P_n(w) \tag{13}$$

In these equations, A_n, B_n, C_n, D_n, F_{1n}, and F_{2n} are constants to be determined from boundary conditions; x_∞ and T_∞ are ambient values of temperature and mole fraction; T_{i1} and T_{i2} are the temperatures inside drops 1 and 2; and $P_n(w)$ are the Legendre polynomials with $w = \cos\eta$, and $f = \cosh(\mu) - w$.

Profiles of vapor density and temperature have been computed for two juxtaposed water drops by Carstens et al. (4) with comments by Williams and Carstens (5). The formalism is herein generalized, in accordance with Williams (6), to include drops with distinct thermal conductivities and latent heats; this allows consideration of the solid phase as well as the possibility that one sphere is nonvolatile.

Consider next the boundary conditions that fix the above constants. Neglecting temperature jump at the liquid–gas interface (7), we must have at the drop surface:

$$T_e|_{\text{surface}} = T_i|_{\text{surface}} \tag{14}$$

in which the subscripts e and i refer to fields exterior and interior to the drops.

For a second boundary condition, we consider two distinct possibilities: the interaction of a volatile and inert sphere and the interaction of two volatile spheres. The boundary condition characterizing the inert sphere is:

$$\nabla x|_{\text{surface}} \cdot \vec{n} = 0 \tag{15}$$

in which $\vec{n}$ is a unit vector normal to the surface. The condition characterizing the volatile sphere is given as a linearization of the equilibrium condition at the surface (vapor saturation). We write this as follows:

$$x|_{\text{surface}} = b_i T|_{\text{surface}} + c_i \tag{16}$$

in which i identifies the volatile sphere, and b_i and c_i are fitted constants. Thus, if sphere 1 is volatile and sphere 2 inert, there is one set of constants b_1 and c_1. This likewise holds if both spheres are either liquid or solid. If the spheres are of different phases, then i = 1, 2, and the condition of Equation 15 does not apply.

Another independent surface condition is secured by requiring continuity of the energy flux across the liquid–vapor interface;

$$\vec{n} \cdot [L_i cD\nabla x + K_e \nabla T_e - K_i \nabla T_i]_{\text{surface}} = 0 \tag{17}$$

Here again, if sphere 2 is inert, $L_2 = 0$, and K_2 refers to the thermal conductivity of the substance. Two such conditions hold if there are two phases; one if there is one phase.

We must finally add a condition explicitly concerned with the steady-state conservation of energy. For either drop over the surface:

$$\iint (L_i Dc\nabla x + K_e \nabla T)|_{\text{surface}} \cdot \vec{ds} = 0 \tag{18}$$

in which i = 1, 2 and in which $\vec{ds}$ is an element of surface.

The above conditions suffice to specify the constants in Equations 10–13. By taking 10 terms in each series, Williams (7) obtained accuracy to three places for water, ice, and air under low supersaturations (1.05) and near standard atmospheric conditions. This was adequate for certain meteorological applications. The set of linear equations in the constants A_n, B_n, etc., are routinely solved on the computer.

In the important case in which the two drops are volatile and the same liquid, the interior temperatures are uniform. This leads to the simplified solutions:

$$T(\mu,\eta) = (T_s - T_\infty) \Phi + T_\infty \tag{19}$$

and

$$x(\mu,\eta) = (x_s - x_\infty) \Phi + x_\infty \tag{20}$$

in which

$$\Phi = (2f)^{1/2} \sum_{n=0}^{\infty} \{\sinh(j\mu_2) \exp j(\mu - \mu_1)$$

$$- \sinh(j\mu_1) \exp j(\mu_2 - \mu)\} P_n(w) / \sinh[j(\mu_2 - \mu_1)] \tag{21}$$

and in which $j = n + 1/2$. In the above solution, both x_s and T_s are surface values, and T_s represents the temperature of the drop. These values are the same as would be obtained if the drops were isolated. For example, for an isolated drop, if we retain the linearized saturation condition given by Equation 16 and write the single drop power balance:

$$cLD \frac{\partial x}{\partial r}\Big|_{\text{surface}} = K \frac{\partial T}{\partial r}\Big|_{\text{surface}} \ (r = a) \tag{22}$$

that is:

$$(x_s - x_\infty)/(T_\infty - T_s) = (K/cLD) = \Gamma \tag{23}$$

then, these two equations yield:

$$x_s = x_\infty - [x_\infty - (b_1 T_\infty + c_1)/(1 + b_1/\Gamma)] \tag{24}$$

This circumstance arises when the impressed field on a drop satisfies the steady state condition:

$$DLc\nabla x + K_e \nabla T_e = 0$$

where L is the latent heat of the drop liquid. This being the case, it follows that if we imagine x to be unity inside the drop, then $\nabla x = 0$ so that $\nabla T = 0$, and T is constant inside the drop.

Results and Discussion

The evaluation of mass rates of either growth or evaporation is straightforward, and reference is made to Carstens et al. (4) for detailed formulas. The more general treatment presented here does not substantially alter our previous conclusions. The magnitude of relative fall rates in the size regime in which continuum transport is valid tends to restrict our attention to drops of about the same size. Thus for a 10% reduction in growth/evaporation, a separation distance of about 10 times the average radius is required. This interaction can cause some percentage (those paired) of drops to lag their isolated counterparts. The magnitude of such interaction of course depends upon the sizes involved and the total drop population. The brevity of interaction between different sized drops tends to minimize size distribution spreading resulting from this particular effect.

For illustrative purposes, we exhibit temperature and vapor density profiles along the line connecting sphere centers for two water drops growing under a supersaturation ratio of 1.05 at atmospheric pressure and 27°C ambient temperature; these profiles are shown in Figures 1 and 2. Also in Figures 3 and 4, the profiles for a water drop and an inert sphere are shown. It may be of some interest to use the present generalized formalism to treat thermophoretic and diffusiophoretic forces between spheres.

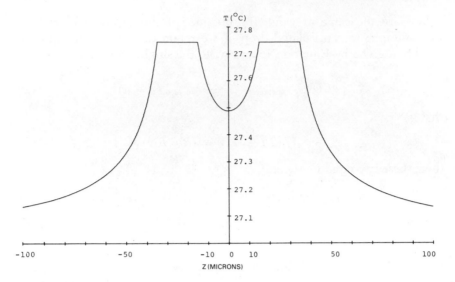

Figure 1.   Temperature vs. distance (Z) along line-of-centers for two water drops of equal radii (20 μ). Supersaturation ratio, 1.05; ambient temperature, 27°C.

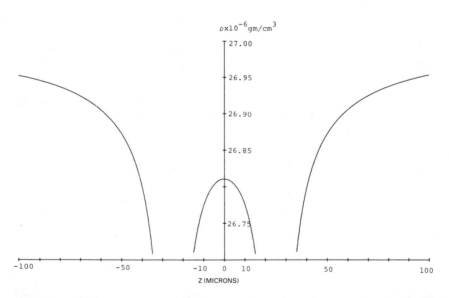

Figure 2.   Vapor density vs. distance (Z) along line-of-centers for two water drops of equal radii (20 μ). Supersaturation ratio, 1.05; ambient temperature, 27°C.

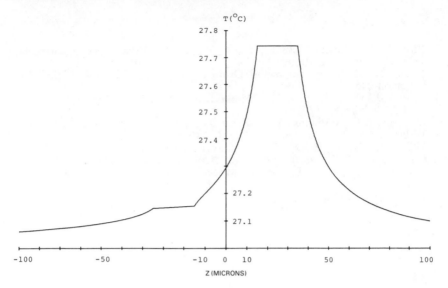

Figure 3. Temperature vs. distance (Z) along line-of-centers for water drop (right) and inert sphere. Supersaturation ratio, 1.05; ambient temperature, 27°C.

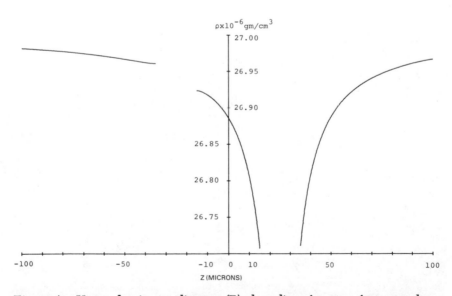

Figure 4. Vapor density vs. distance (Z) along line-of-centers for water drop (right) and inert sphere. Supersaturation ratio, 1.05; ambient temperature, 27°C.

Literature Cited

1. Bird, R. B., Stewart, W. E., Lightfoot, E. N., "Transport Phenomena," Wiley, New York, 1960.
2. Fuchs, N. S., "Evaporation and Droplet Growth in Gaseous Media," Pergamon, New York, 1959.
3. Morse, P. M., Feshbach, H., "Methods of Theoretical Physics," McGraw Hill, New York, 1953.
4. Carstens, J. C., Williams, A. L., Zung, J. T., *J. Atmos. Sci.* (1970) **27**, 798.
5. Williams, A. L., Carstens, J. C., *J. Atmos. Sci.* (1971) **28**, 1298.
6. Williams, A. L., Ph.D. Thesis, University of Missouri—Rolla, 1972.
7. Carstens, J. C., Carter J., "Proc. Int. Colloquium on Drop and Bubbles," D. J. Collins, M. S. Plesset, M. M. Saffren, Eds., p. 529, Cal Tech.-JPL, 1974.

RECEIVED December 29, 1976.

"Group" Combustion of Droplets in Fuel Clouds. I. Quasi-steady Predictions

M. LABOWSKY and D. E. ROSNER

Chemical Engineering Section, Yale University, Department of Engineering and Applied Science, New Haven, CT 06520

Conditions are established for "group" combustion (multiple particle burning with a common flame) for a quasi-steadily burning fuel droplet cloud. Two conditions are investigated: incipient group combustion, i.e., the onset of flame sharing by particles at the cloud center, and total group combustion, i.e., flame sharing by the entire cloud of fuel particles. The corresponding quantitative conditions are established using two different approaches: a continuum method in which the effectiveness of oxidizer penetration into the cloud is computed (in analogy to reactant diffusion into a porous catalyst pellet) and a superposition (image) method in which flame locations for particle arrays are calculated. The results indicate that quasi-steady fuel clouds with interparticle spacings of practical interest will burn in the group mode.

It has been commonly assumed in the past that during spray combustion each droplet is surrounded by its own envelope flame (*1*). However, quantitative criteria for when particles burn individually, as opposed to when they pool their vapor and burn as a group, have apparently never been stated clearly (*2*), even for simple geometric circumstances. Indeed, recent experimental work on probing burning sprays suggests that individual droplet flames are encountered rarely in conventional spray combustors (*3, 4*). Of course, such observations do not imply that the individual flame burning concept is incorrect per se—only that in the combustors investigated local operating conditions precluded the individual droplet burning mode.

This chapter is the first part of a series on the development of quantitative criteria for predicting in which mode a fuel droplet array will burn.

0-8412-0383-0/78/33-166-063$05.00/0 © 1978 American Chemical Society

The concepts used and the resulting criteria should ultimately help in the understanding and design of spray combustors. However, to gain insight into this general class of problems, we do not consider fuel sprays here, but rather the simpler case of fuel clouds. In particular, we ask: if a quiescent fuel droplet cloud were to approach a quasi-steady (QS) state, under what conditions would it burn as a group while in that state? That is, we seek group combustion criteria based on a quasi-stationary burning cloud model.

A QS cloud is an extension of the well known QS model for a single fuel droplet as discussed in Refs. 1 and 5. Thus, in this chapter, initial and final transients, as well as free convection effects, are neglected, and the particle and cloud radii are considered to be slowly varying functions of time. A faithful experimental analog of a QS cloud would be a burning array of fixed spherical porous pellets through which enough fuel is fed to maintain steady-state combustion. As will become evident, we are not concerned here so much with immediate application to real spray combustors or with mathematical rigor but rather with gaining useful insight into group behavior by first considering the most tractable cases.

In keeping with these objectives, we obtain two QS group combustion criteria, one which we call incipient group combustion (when only the droplets in the center of the cloud burn as a group) and the other which we call total group combustion in which all the droplets burn as a group. To understand better what is meant by incipient and total group combustion, consider Figure 1, which shows actual flame locations calculated using a method described below.

When the fuel droplets are very far apart, they burn as separate entities, each surrounded by its own envelope flame (Figure 1a). When they are moved closer together, or as more particles are added to the cloud, oxygen has more difficulty in penetrating the cloud by diffusion, and as a result the flame radii of the interior particles increases. When the gaseous oxidizer has been depleted sufficiently, the flames around the droplets in the center of the cloud will just touch, and these particles will be considered to burn as a group (Figure 1b) while other particles continue to burn individually. We call this situation incipient group combustion. With further reductions in particle separation, additional particles join the group until all fuel particles in the cloud ultimately burn with a common flame (Figure 1d). This situation we call total group combustion.

We establish QS group combustion criteria using two different approaches. In the first (in the following two sections), the cloud is treated as a continuum in which the cloud droplets act as distributed sources of fuel and sinks of oxidizer. We base our group combustion criteria on the ability of ambient oxidizer to penetrate the cloud by diffusion, drawing

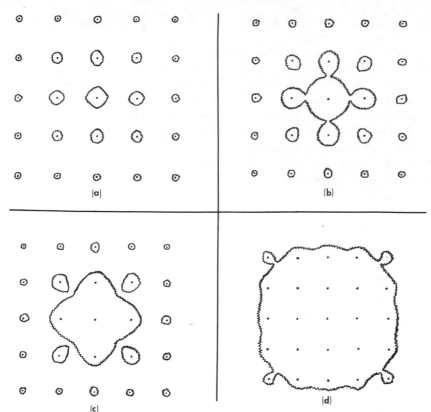

Figure 1. Superposition calculations of flame locations as a function of par-
ticle separation for a dodecane fuel cloud consisting of 125 spherical particles
burning in air. Incipient group combustion is shown in (b), and total group
combustion is shown in (d).

on analogies to the classical chemical engineering problem of reactant
penetration into porous catalyst pellets. In the second approach ("Superposition Calculation of Flame Locations"), we calculate the flame locations for cubical arrays containing up to 729 particles by transforming the
Schvab–Zeldovich combustion equations to the Laplace equation and
solving by a superposition method, which is an approximation of the
well known method of images. Since the continuum approach is rigorously valid only when there is a large number of particles in the cloud
and since the superposition method is feasible only when the number is
small, we establish conditions valid over the entire range of cloud sizes
by combining the results from these two approaches.

In the last section, our QS criteria for group combustion are applied
to the case of a dodecane cloud burning in air. Results obtained indicate
that virtually all QS burning clouds of practical interest would burn as a
total group.

QS-*Continuum Criterion for Incipient Group Combustion*

Approach. In this section we assume either that after ignition the fuel cloud goes through a period of initial transients during which any premixed (inter droplet) oxidizer is consumed or that ignition follows oxidant penetration into an initially oxidant-free fuel droplet array. After a certain period, such a cloud would settle into a quasi-steady state during which whatever oxidizer is needed to maintain combustion is provided by inward diffusion from the ambient. We then ask, under what conditions will the burning fuel droplets constitute sufficiently strong oxidizer sinks that oxidizer is unable to reach the cloud center? That is, under what conditions will it be impossible to maintain individual flame droplet combustion at the cloud center? (Alternatively, one can calculate the critical cloud parameters based on the condition of droplet flame touching caused by local oxidizer depletion at the center of the cloud. Results of such flame-touch calculations were virtually indistinguishable from the present, simpler condition.)

Analysis. To answer the question posed above, we treat the fuel cloud as a pseudo homogeneous or continuum phase $(6, 7)$ with the individually burning particles treated as point sinks of oxidizer. Such an approach is rigorously defensible when the cloud is sufficiently dilute that:

$$R_c \gg L \gg R_f > R_p \gg l \qquad (1)$$

where R_c, L, R_f, and R_p are, respectively, the characteristic cloud dimension, interparticle separation, particle flame radius, and particle radius. Further, if:

$$(\rho/\rho_p) \cdot \ln(1 + B_{comb}) \ll 1 \qquad (2)$$

then QS single particle theory can be applied $(8, 9)$, with the result that each particle has an oxidizer sink strength of $(1, 5)$:

$$\nu \dot{m}_p = 4\pi \overline{\rho D}_F \nu R_p \ln(1 + B_{comb}) \qquad (3)$$

where ν is the stoichiometric oxidizer/fuel mass ratio and:

$$B_{comb} = \frac{\omega_{F,p} + (\omega_O/\nu)}{1 - \omega_{F,p}} \qquad (4)$$

$\omega_{F,p}$ is the mass fraction of fuel at the particle's surface, where thermodynamic equilibrium is assumed to exist, and ω_O is the ambient oxidizer mass fraction "seen" by a droplet. Provided the inequalities in Equation 1 are satisfied, ω_O is very nearly the local phase volume-averaged oxidizer mass fraction and is treated as such in this chapter.

To establish a QS incipient group combustion criterion, the oxidizer and total gas-phase continuity equation must be solved for the oxidizer concentration (in the volume-averaged sense). In so doing, we found that the condition for incipient group combustion corresponded to such a dilute spherical cloud that certain terms in most general conservation equations were of negligible importance. (We solved these equations formally by using several methods, including weighted residuals (*10*) and Frobenius series methods. Results obtained by considering Equation 5, which neglects the effects of Stefan flow except at the particles' surfaces, were within about 5% of those obtained by considering all of the neglected terms for the case of dodecane burning in air. This simplification is therefore consistent with our goal of gaining insight into group behavior.) Under conditions of incipient group combustion, therefore, we need consider only the simplified (radial convection-free) oxidizer conservation equation:

$$\frac{1}{r^2}\frac{\mathrm{d}}{\mathrm{d}r}\left(r^2\overline{\rho D}\,\frac{\mathrm{d}\omega_O}{\mathrm{d}r}\right) \cong \nu\dot{m}_p N_p \tag{5}$$

with $N_p = 0$ for $r > R_c$.

When $\omega_O/\nu \ll 1$, as for the case of most hydrocarbons burning in air, the dimensionless driving force for particle combustion can be linearized as:

$$\ln(1 + B_{comb}) \cong \omega_{F,p,eff} + (\omega_O/\nu) \tag{6}$$

where $\omega_{F,p,eff}$ is the effective fuel mass fraction at a particle's surface. Explicitly:

$$\omega_{F,p,eff} \equiv \ln(1 - \omega_{F,p})^{-1} \tag{7}$$

Thus, the pseudo-homogeneous sink term in Equation 5 becomes the sum of a zero-order and first-order contribution. If $\omega_{F,p}$ is constant throughout the cloud (that is, if all particles have nearly the same temperature), then inside the cloud ($0 < r < R_c$) Equation 5 can be written as:

$$\frac{1}{r^2}\frac{\mathrm{d}}{\mathrm{d}r}\left(r^2\overline{\rho D}_O\,\frac{\mathrm{d}c_*}{\mathrm{d}r}\right) = 4\pi\overline{\rho D}_F R_p N_p c_* \tag{8}$$

where:

$$c_* \equiv \nu\omega_{F,p,eff} + \omega_O \tag{9}$$

In assuming $\omega_{F,p}$ to be a constant, we anticipate a heat/mass compensation effect which tends to suppress particle temperature variation through a QS cloud. Specifically, the decreased oxidizer concentration found as one penetrates the cloud, which tends to decrease particle temperature, is offset by an increase in the local ambient temperature. (Such an effect has been observed recently for the case of evaporation of interacting particles of low volatility (11). Extension of these results follows, since for the conditions of incipient group combustion, the variable $T +$ $[Q/(\nu c_p)]\omega_O$ satisfies the Laplace equation both inside and outside the cloud. The condition for total heat/mass compensation follows provided that $Q >> L$).

Cast in the form of Equation 8, the problem is equivalent to the classical chemical engineering problem of determining the effectiveness of reactant penetration into a porous catalyst pellet which is the site of a distributed pseudo-homogeneous first-order reaction. We can then quote well known results (12, 13). Inside the cloud c_* is governed by:

$$c_* = c_*\big|_{R_c} \cdot \frac{\sinh(\psi r/R_c)}{\sinh(\psi)} \cdot \frac{R_c}{r} \qquad (0 < r < R_c) \qquad (10)$$

where ψ, the cloud Thiele modulus, is simply the ratio of the square root of the characteristic time for the reactant to diffuse into the cloud to the characteristic reaction time. Alternatively, ψ can be regarded as the ratio of the overall radius to the characteristic reactant penetration depth. The smaller the value of ψ, therefore, the easier it is for the reactant to penetrate the cloud. Assuming that $\overline{\rho D} \approx \overline{\rho D_F}$, the Thiele modulus can be written:

$$\psi = (3\phi)^{1/2} (R_c/R_p) \qquad (11)$$

where ϕ is the dispersed phase (droplet) volume fraction.

Rearranging Equation 10 we obtain an explicit expression for the oxidizer mass fraction inside the cloud ($r < R_c$):

$$\omega_{O,int} = \left(\omega_O\big|_{R_c} + \nu\omega_{F,p,eff}\right) \frac{\sinh[\psi r/R_c]}{\sinh(\psi)} \cdot \frac{R_c}{r} - \nu\omega_{F,p,eff} \qquad (12)$$

But from Equation 5 the oxidizer mass fraction ω_O satisfies the Laplace equation outside of the cloud. Therefore:

$$\omega_{O,ext} = \left(\omega_O\big|_{R_c} - \omega_{O,\infty}\right)(R_c/r) + \omega_{O,\infty} \qquad (r > R_c) \qquad (13)$$

Since conservation of mass requires that $d\omega_{0,int}/dr = d\omega_{0,ext}/dr$ at $r = R_c$, it can be shown that:

$$\omega_0\Big|_{R_c} = \frac{\omega_{0,\infty} - \nu\omega_{F,p,eff}\,(\psi\coth(\psi) - 1)}{\psi\coth(\psi)} \tag{14}$$

which, upon insertion into Equation 12, completes the description of $\omega_0(r)$.

Results. Figure 2 shows the results of the above calculations for the case of a spherical dodecane cloud burning in air. Here the local oxygen mass fraction, normalized by its ambient value, is plotted against the distance from the cloud center, measured in cloud radii, with ψ as a parameter. Notice that when $\psi = 0$ the cloud offers no resistance to oxygen penetration and the oxidizer mass fraction is at the ambient value every-

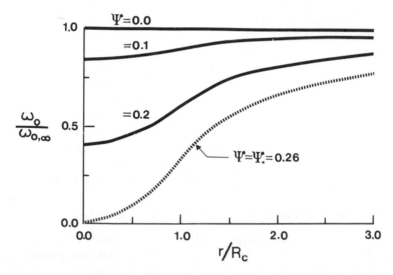

Figure 2. Oxygen concentration profiles as a function of the dimensionless oxidizer sink strength ψ for spherical dodecane fuel clouds burning in air. Droplet cloud boundary at $r/R_c = 1.0$.

where. Increasing ψ slightly, however, decreases significantly the amount of oxygen reaching the cloud center, indicating that the burning particles act as very effective oxygen sinks. To determine the incipient group combustion condition, we increase ψ until, at some critical value denoted by ψ_*, $\omega_0 = 0$ at the cloud center. At this condition, individual envelope flame combustion is certainly impossible at the cloud center, and therefore particles near the cloud center must burn in a group mode, even though particles away from the center may burn individually. The oxygen profile corresponding to this critical Thiele modulus ψ_* is shown by the broken line in Figure 2. ψ_* can be found formally by imposing the condition $\displaystyle\lim_{r \to 0}\ \omega_{0,\text{int}}(r) \to 0$. In so doing, Equations 12 and 13 can be rearranged to give:

$$\psi_* = \cosh^{-1}\left\{ \frac{\omega_{0,\infty}}{\nu\omega_{F,p,\text{eff}}} + 1 \right\} \qquad (15)$$

Since ψ_* is independent of R_c and ϕ, Equation 11 implies that the critical cloud size for which incipient group combustion will occur, $(R_c/R_p)_*$, scales with the inverse square root of the droplet volume fraction, i.e.:

$$(R_c/R_p)_* = \psi_*(3\phi)^{-1/2} \qquad (16)$$

Thus, for a given ϕ, if the actual cloud radius (R_c/R_p) is less than the predicted critical value $(R_c/R_p)_*$, then all of the particles in a quasi-stationary burning could will be able to burn individually.

For most hydrocarbons burning in air, $\omega_{0,\infty}/\nu$ is only about 0.07, and $\omega_{F,\text{eff}}$ is about 2, so that the pseudo-homogeneous source term in Equation 6 depends only weakly on ω_0. For this case, ψ plays the role of a zero-order Thiele modulus (13). Thus, the present continuum analysis clearly reveals that the dimensionless parameter which dictates the onset of incipient group combustion is simply a Damköhler number (Thiele modulus) which takes on a critical value (Equation 15) dependent only on the ambient oxidizer mass fraction, on the fuel vapor mass fraction at the droplet surface, and on the stoichiometric oxidizer/fuel mass ratio.

QS-Continuum Criterion for Total Group Combustion

Approach. The continuum total group combustion criterion is established by asking: when will the cloud burn as a large pseudo-droplet with the flame located just outside of the cloud droplet region? That is, when will the evaporating particles inside the cloud provide sufficient vapor so that fuel and oxidizer mix in stoichiometric amounts at the cloud boundary?

Analysis. To answer this question formally the fuel species and oxidizer species conservation equations would have to be solved. But again, we found a posteriori that the condition for group combustion corresponded to such a dilute cloud that these continuity equations could be approximated accurately enough by neglecting the terms involving Stefan flow convection. Therefore:

$$\frac{1}{r^2}\frac{d}{dr}\left(r^2\rho\overline{D}_F\,\frac{d\omega_F}{dr} \right) = -N_p\dot{m}_p \qquad (r < R_c) \qquad (17)$$

$$\frac{1}{r^2}\frac{d}{dr}\left(r^2\overline{\rho D}_0\frac{d\omega_0}{dr}\right)=0 \qquad (r>R_c) \qquad (18)$$

Since the particles inside the cloud are evaporating:

$$\dot{m}_p = 4\pi\overline{\rho D}_F R_p \ln(1+B_{vap})$$

$$= 4\pi\overline{\rho D}_F R_p \ln\left(\frac{1-\omega_F}{1-\omega_{F,p}}\right), \text{ or}$$

$$\dot{m}_p \approx 4\pi\overline{\rho D}_F R_p[\omega_{F,p,eff}-\omega_F] \qquad (19)$$

where terms of order $\omega_F{}^2$ have been neglected. We again anticipate a heat/mass compensation effect so that the particle temperatures, and thus $\omega_{F,p}$, are treated as constants through the cloud. Using this linearized rate law and introducing the mass fraction c_{**}:

$$c_{**} \equiv [\omega_{F,p,eff}-\omega_F]/\omega_{F,p,eff} \qquad (20)$$

the fuel continuity equation reduces to a form similar to Equation 9:

$$\frac{1}{r^2}\frac{d}{dr}\left(r^2\overline{\rho D}_F\frac{dc_{**}}{dr}\right)=4\pi\overline{\rho D}_F R_p N_p c_{**} \qquad (21)$$

Again drawing an analogy with transport through a catalyst pellet, the fuel flux at the cloud surface, $(d\omega_F/dr)$ at $r=R_c$, can be written in terms of a Thiele modulus:

$$\Phi = (3\phi)^{1/2}(R_c/R_p) \qquad (22)$$

and an "effectiveness factor." (An effectiveness factor in catalysis is the ratio of the actual reaction rate to that rate which the pellet would experience if the reaction were not pore diffusion-limited.):

$$\eta = \frac{3}{\Phi}\left[\frac{1}{\tanh(\Phi)}-\frac{1}{\Phi}\right] \qquad (23)$$

Explicitly,

$$\overline{\rho D}_F\frac{d\omega_F}{dr}\bigg|_{r=R_c} = \frac{\overline{\rho D}_F\omega_{F,p,eff}}{R_c}\cdot 3\Phi^2\eta \qquad (24)$$

The functional form of Φ is the same as that of ψ in "Incipient Group Combustion," but these two Thiele moduli will, in general, have different numerical values. Since the total group combustion criteria requires that the fuel vapor mass flux equals the stoichiometric oxidizer flux at the cloud surface:

$$\rho D_F \frac{d\omega_F}{dr}\bigg|_{r=R_c} = \frac{\rho D_O}{\nu}\frac{d\omega_O}{dr}\bigg|_{r=R_c} = \frac{\overline{\rho D_O \omega_{O,\infty}}}{R_c \nu} \qquad (25)$$

The total group combustion Thiele modulus, Φ_{**}, satisfies the transcendental equation:

$$3\Phi^2_{**}\eta(\Phi_{**})\overline{\rho D_F \omega_{F,p,eff}}/R_c = \overline{\rho D_O \omega_{O,\infty}}/(\nu R_c) \qquad (26)$$

where $\eta(\Phi)$ is given by Equation 23. The critical cloud radius for total group combustion $(R_c/R_p)_{**}$ is then simply:

$$(R_c/R_p)_{**} = (3\phi)^{-1/2}\,\Phi_{**} \qquad (27)$$

Thus, for a given fuel volume fraction ϕ, if the actual cloud radius (R_c/R_p) is greater than $(R_c/R_p)_{**}$ (Equation 27), then a quasi-stationary cloud will burn with a flame surrounding the entire cloud. (Only in the limit $\eta \to 1$ is the Suzuki–Chiu criterion of the same form as Equation 26; moreover, their criterion for total group combustion contains the a priori unknown fuel vapor mass fraction at the droplet cloud center. Ironically, in their numerical estimates they put $\omega_F(0) = 1$; an ad hoc assumption which leads to a total group combustion condition corresponding to a cloud even more dilute than that predicted by Equation 26.) Again, our analysis shows quite clearly that the dimensionless parameter governing the phenomenon of total group combustion is simply a cloud Damköhler number (Thiele modulus), the critical value of which is a calculable function (Equation 26) of $[\rho D_O/\rho D_F] \cdot [\omega_{O,\infty}/\nu\omega_{F,p,eff}]$.

Superposition Calculation of Flame Locations

Approach. The use of continuum models is defensible only when the particles are sufficiently far apart that suitable averaged oxidizer and fuel mass fraction can be defined. This implies not only the $R_p/L << 1$ but also the $R_f/L << 1$, where R_f is the individual envelope flame radius, and L is the interparticle distance. Indeed, the use of a continuum approach to establish the incipient group combustion criterion may be suspect since this last inequality is not met. Further, continuum models cannot be defended when the number of particles in the cloud is small $(\mathcal{N} < 1000)$. In the light of these possible limitations on continuum methods, an alternative approach was deemed desirable so that the continuum incipient group combustion criterion can be checked and so that continuum results can be extended into the small cloud region. In such an approach we calculate the actual flame locations using the superposition method which is also (less correctly) called the method of images.

Analysis. In the region between the fuel droplets, when $D_O \approx D_F$, the fuel and oxidizer continuity equations can be combined and expressed in the Schwab–Zeldovich form $(1, 9)$:

$$\rho \vec{v} \cdot \nabla [\omega_F - (\omega_O/\nu)] = \overline{\rho D} \nabla^2 [\omega_F - (\omega_O/\nu)] \tag{28}$$

Assuming the background gas species (i.e., the gas species which is not fuel, oxidizer, or product) is stagnant and that the reaction is confined to a thin reaction zone, the total mass-averaged gas mixture velocity, $\vec{v}$, can be written:

$$\vec{v} = -\frac{D \nabla [\omega_F - (\omega_O/\nu)]}{1 - \omega_F + (\omega_O/\nu)} \tag{29}$$

Introducing a new transformed dependent variable ξ, defined by

$$\xi = \frac{\ln \left[\dfrac{1 - \omega_F + (\omega_O/\nu)}{1 + (\omega_{O,\infty}/\nu)} \right]}{\ln \left[\dfrac{1 - \omega_{F,p}}{1 + (\omega_{O,\infty}/\nu)} \right]} \tag{30}$$

it can be seen from a combination of Equations 28 and 29 that ξ satisfies Laplace's equation:

$$\nabla^2 \xi = 0 \tag{31}$$

which must be solved subject to the boundary conditions $\xi = 1$ on all particle surfaces and $\xi = 0$ far from the array. (We have assumed a compensation effect again. If ξ assumes different values on different particles, an analysis similar to that in Ref. *11* would have to be performed. Indeed, this approach is currently being used to ascertain to what extent, if any, the compensation effect is invalid.) Cast in this form the problem of determining the mass fraction fields between interacting burning particles is reduced to one similar to finding the electrical field between charged conductors and is amenable to solution by the method of images. This method, which has recently been extended to the multiparticle case (14), provides a means for locating mass point sources and sinks so that Equation 31 and the boundary conditions are satisfied. Using this method to find the ξ-field, the flame will be located on the locus of points where ω_O, $\omega_F = 0$, i.e., where:

$$\xi = \xi_f = \frac{\ln \{[1 + (\omega_{0,\infty}/\nu)]^{-1}\}}{\ln \{[1 - \omega_{F,p}]/[1 + (\omega_{0,\infty}/\nu)]\}} \qquad (32)$$

Quasi-steady theory predicts isolated particle flame radii which are greater than those experimentally observed. While a discussion of causes of this descrepancy is beyond the scope of this chapter, the reader should keep in mind that absolute errors are associated with equations such as Equation 32. Nevertheless the functional dependency on ambient oxidizer concentration is of principal interest here.

An interesting and necessary aspect of this ξ-transformation is that the classical QS single particle results are obtained when only one particle is considered. As a consequence, in using this transformation to solve the multi-particle combustion problem, we need not invoke any further assumptions than those commonly used in the single particle droplet combustion problem. Further, the images method is a fairly general one for treating interacting particle evaporation and combustion problems. Using this method, interacting particles of any relative size, in any geometric arrangement, and with different chemical composition can be considered (11). Here, however, we restrict our consideration to equal sized particles in cubical array, with each particle separated from its nearest neighbors by a center-to-center distance, $\mathcal{L}$. Referring again to Figure 1, to obtain the incipient group combustion point for a given array, we adjust $\mathcal{L}$ so that the flames of the particles at the center of the array just touch (Figure 1b). The total group combustion point is determined by adjusting $\mathcal{L}$ so that the flames around the particles at the corner of the cubical array just touch the flames of their nearest neighbors.

In general, when using the method of images, the number of point sources which must be considered in order to estimate the ξ-field increases rapidly as the number of interacting particle increases. Computational costs usually become prohibitive when the number of particles approaches 20. However, since group combustion occurs at extremely large inter-particle separations, the ξ-fields can be approximated by using only the primary sources in the image series. That is, the ξ-fields can be found by treating the particles as point sources and by simply superimposing their respective fields. (This point source approximation is valid provided $\mathcal{N}/\mathcal{L} << 1$, where $\mathcal{L}$ is the average dimensionless distance of all particles from the center particle. For the case of 729 dodecane particles, $\mathcal{L}_* \approx O(10^4)$ so that the error in using this approximation for this case is less than about 10%.) Using this point source approximation, arrays consisting of up to 729 particles could be considered readily. Since it would be misleading to say that we used the method of images to consider such large arrays, we will hereafter refer to this method simply as a superposition method.

Discussion

The results obtained from the continuum and superposition calculations are compared in Figure 3 for the case of a dodecane cloud burning in air. Here the logarithm of the number of particles is plotted against the logarithm of the interparticle separation. Since $\mathcal{N}$ and $\mathcal{L}$ are related to R_c and ϕ by:

$$\mathcal{N} = \phi (R_c/R_p)^3 \tag{33}$$

$$\mathcal{L} \cong (4.19/\phi)^{1/3} \tag{34}$$

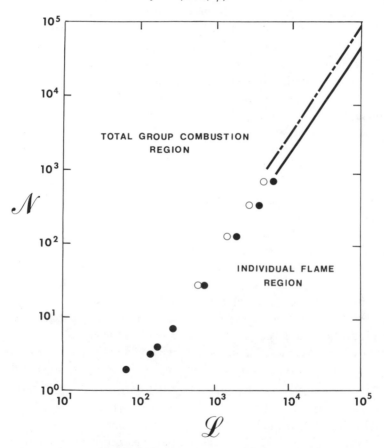

Figure 3.   *Criteria for the quasi-steady group combustion of a spherical dodecane fuel droplet cloud in air.* (——) *QS incipient group combustion criterion* (see "*Incipient Group Combustion*"); (— — —) *QS total group combustion criterion* (see "*Total Group Combustion*"); (●) *superposition incipient group combustion criterion;* (○) *superposition total group combustion condition* (see "*Calculation of Flame Locations*").

an inverse square root dependence of R_c on ϕ is reflected in a 3/2-power dependence of $\mathcal{N}$ on $\mathcal{L}$. The broken line represents the continuum total group combustion criterion, and the solid line represents the incipient group combustion criterion. If a cloud were to approach a steady state located to the right of the broken line, it would burn in a single particle mode while in that state. If the steady state were to fall to the left of the solid line, the cloud would burn entirely as a group. If it were to fall somewhere in between these two lines, the cloud would burn in some mixed mode.

The continuum results are truncated at $\mathcal{N} = 1000$, since this method is not rigorously applicable to clouds with fewer particles. For such clouds, the superposition method must be used. In Figure 3, the data represent the superposition incipient group combustion points, and the circles represent the total group combustion points. These points compared very well with the continuum results, except that they are shifted slightly to the right. This may be a geometric effect, since the continuum calculations were applied to a cubical array. The only substantial differences between these calculations occur when $\mathcal{N} < 10$. If the continuum lines were continued into this region, their slopes would not change. The superposition calculations, however, show the proper merging of the incipient and total group combustion points when a cloud consists of a single particle surrounded by its nearest neighbors. For this case, the incipient and total group combustion conditions are, of course, the same. It is somewhat surprising that the 3/2-power dependence of $\mathcal{N}$ on $\mathcal{L}$, as predicted by the continuum calculations, continues well into the region where this method is no longer rigorously valid. Since it does, our initial concern that the continuum incipient group combustion criterion is not rigorously defensible when R_f is of the order of $L/2$ seems to be needless. The reason for this may be that the particle flame radii become large only as L approaches the incipient group combustion separation, so that in fact R_f is much less than L except near L_*. This conjecture is supported by the superposition calculations.

The quantitative results contained in Figure 3 are quite instructive. For example, if a cloud containing only 10^3 particles were to burn in a QS-individual flame mode, it would have to have a mean interparticle separation of about 7000 R_p. By comparison, droplets in a spray combustor typically have interparticle separations of $(10–100)R_p$, implying that QS clouds with particle spacings of practical interest would never burn in the individual flame mode. This results from the remarkable efficiency of such cloud particles in preventing oxidizer penetration by diffusion into the cloud.

While it is tempting, it would be premature to apply these equations and findings directly to more complex spray combustor situations. Apart from obvious differences in overall geometry, in practical sprays three effects are superimposed: transients associated with oxidizer entrained in the fuel injector region, droplet-size-dependent relative motion between the fuel droplets and the surrounding gas, and oxidizer and product transport by turbulence and convection. Rather, our present QS and future transient studies of the behavior of quiescent fuel droplet clouds should be viewed as necessary first steps in the qualitative and quantitative theoretical understanding of fuel droplet sprays. Future work should be concerned not only with the conditions under which theoretical group combustion occurs in fuel sprays but also with the implications of such cooperative phenomena for combustion efficiency in volume-limited systems, and pollutant emissions.

Conclusions

To shed light on the validity of the envelope flame concept when applied to multidroplet fuel arrays, we have considered quantitatively the behavior of fuel droplet clouds in the tractable limiting case of quiescent, quasi-steady combustion. Our analysis clearly reveals that the group combustion phenomenon is governed by a Damköhler ratio (Thiele modulus) identical in form to that previously used in the treatment of simultaneous diffusion and chemical reaction within porous media. In the light of the quantitative formulas developed and the numerical results reported here for the case of dodecane burning in quiescent air, we find that the diffusive mode of oxidizer transport is inadequate to prevent group combustion for interdroplet spacings and cloud sizes of practical interest. In work currently in progress, we are investigating the efficiency of other modes of oxidizer transport (e.g., turbulence) and the effects of transient (non-QS) behavior.

While complicated by such additional phenomena as turbulence, droplet slip, etc., qualitatively similar phenomena should occur locally in fuel droplet sprays of practical interest. Our hope is that the insights gained in the present theoretical treatment of quiescent fuel droplet arrays ultimately prove useful in the understanding and mathematical modeling/optimal design of practical spray combustors.

Nomenclature

B_{comb} = mass transfer driving force with combustion
B_{vap} = mass transfer driving force without combustion
c = effective normalized reactant concentration
c_p = specific heat (per g of mixture)

D_i = Fick diffusivity for species i relative to mixture

l = molecular mean-free-path (we assume $R_p >> l$)

L_v = latent heat of fuel vaporization (per g)

L = interparticle separation

$\mathcal{L}$ = dimensionless interparticle separation, L/R_p

Le   = Lewis number (mass/heat diffusivity ratio)

$\dot{m}_p$ = individual particle gasification rate

N_p = fuel particle number density

$\mathcal{N}$ = total number of particles in cloud

p = total pressure

Q = heat released per unit mass of fuel vapor combusted

QS   = quasi-steady (see Ref. 9)

r = radial coordinate (from cloud center (line))

R_c = fuel cloud radius

R_p = fuel particle radius

R_f = envelope flame radius

T = absolute temperature

v = mass average fluid velocity

η = effectiveness factor

ν = stoichiometric oxidizer/fuel mass ratio

ρ = absolute mass density

ϕ = condensed fuel volume fraction $(4\pi/3)R_p^3 N_p$

Φ = total group combustion Thiele modulus

ψ = incipient group combustion Thiele modulus

ω_i = mass fraction of species i ($i = $ O,F)

$\omega_{F,p,eff}$ = effective fuel mass fraction at the particle surface, Equation 7

ξ = dependent variable defined by Equation (30)

Subscripts

f   = at diffusion flame location

F   = pertaining to fuel

O   = pertaining to oxidizer

p   = pertaining to (or in equilibrium with) a single fuel particle

*   = pertaining to incipient group combustion condition

**   = pertaining to total group combustion condition

int   = interior (of cloud)

ext   = exterior (of cloud)

∞ = pertaining to conditions far from cloud

Miscellaneous

∇ = spatial gradient operator

∇^2 = Laplacian differential operator

$\overline{(\ \ )}$ = average value of ( )

Literature Cited

1. Williams, F. A., "Combustion Theory," Addison-Wesley, Reading, Mass., 1965.
2. Suzuki, T., Chiu, H. H., "Multi-Droplet Combustion of Liquid Propellants," *Proc. Int. Symp. Space Techol. Sci., 9th, Tokyo,* 1971, 145–154.
3. Chigier, N. A., McCreath, C. G., "Liquid Spray Burning in the Wake of a Stabilizer Disk," *Symp. (Int.) Combust., 14th,* 1973, 1355–1363.
4. Onuma, Y., Ogasawara, M., "Studies on the Structure of a Spray Combustion Flame," *Symp. (Int.) Combust., 15th,* 1975, 453–465.
5. Rosner, D. E., "Liquid Droplet Vaporization and Combustion," in "Liquid Propellant Rocket Motor Combustion Instability," D. T. Harrje, Ed., Section 2.4, pp. 74–100, U.S. Govt. Printing Office, **NASA SP-194,** 1972.
6. Zung, J. T., "Evaporation Rates and Lifetimes of Clouds and Sprays in Air —The Continuum Model," *J. Chem. Phys.* (1967) **47,** 3578–3581.
7. Zung, J. T., "Evaporation Rate and Lifetimes of Clouds III—Effects of Cloud Expansion,, *J. Chem. Phys.* (1968) **48,** 5181.
8. Crespo, A., Liñan, A., "Unsteady Effects in Droplet Evaporation and Combustion," *Combust. Sci. Technol.* (1975) **11,** 9–18.
9. Rosner, D. E., Chang, W. S., "Transient Evaporation and Combustion of a Fuel Droplet Near Its Critical Temperature," *Combust. Sci. Technol.* (1973) **7,** 145.
10. Finlayson, B. A., "The Method of Weighted Residuals and Variational Principles," Vol. 87, "Mathematics in Science and Engineering," Academic, New York, 1972.
11. Labowsky, M., "A Method for Determining the Effects of Chemical Composition on the Heat/Mass Transfer Rates of Interacting Spherical Particles," 69th Annual Meeting of the AIChE, Chicago, 1976.
12. Satterfield, C. N., "Mass Transfer in Heterogeneous Kinetics," M.I.T., Cambridge, 1970.
13. Weekman, V. W., Gorring, R. L., "Influence of Volume Change on Gas-Phase Reactions in Porous Catalysts," *J. Cataly.* (1965) **4,** 260–270.
14. Labowsky, M., "The Effects of Nearest Neighbor Interactions on the Evaporation Rate of Cloud Particles," *Chem. Eng. Sci.* (1976) **31,** 803.

RECEIVED December 29, 1976. Supported in part by AFOSR (Grant 73-2487), the National Science Foundation (Grant GK 25883), and ONR (Project Squid, Grants N00014-67-A-0226-0005, Nr-098-038); presented August 1976.

Experimental Methods

Evaporation and Combustion of Uniformly Sized Hexane Droplets in a Refractory Tube

BYUNG CHOI[1] and STUART W. CHURCHILL

Department of Chemical and Biochemical Engineering, University of Pennsylvania, Philadelphia, PA 19104

NO$_x$ values of 11–24 ppm (as NO) were attained by evaporating and then burning a string of hexane droplets. Uniformly sized and spaced droplets about 230 μm in diameter were produced by vibrating a capillary tube. The droplets were introduced down the axis of a ceramic tube 0.864 m long and 9.54 mm i.d. in a co-current turbulent stream of air. Thermal feedback for evaporation of the droplets and for preheating of the resulting vapor–air mixture occurred by convective heating of the downstream wall by the hot products of combustion, followed by convection to the gas stream and droplet. The exit concentration of NO$_x$ decreased, but CO increased (0.2–0.4%) when the flame front was shifted downstream by changing operating conditions.

Chen and Churchill (1) demonstrated that a flame of premixed air and propane can be stabilized inside a refractory tube. They measured the wall-temperature profile and determined the flashback and blowoff limits as a function of the fuel-to-air ratio for tube diameters of 4.76 and 9.53 mm.

They subsequently (2) developed a one-dimensional mathematical model in the form of coupled differential and integro-differential equations, based on a gross mechanism for the chemical kinetics and on thermal feedback by wall-to-wall radiation, conduction in the tube wall, and convection between the gas stream and the wall. This model yielded results by numerical integration which were in good agreement with the experimental measurements for the 9.53-mm tube. For this tube diameter, the flows of unburned gas for stable flames were in the turbulent regime.

---

[1] Current address: Mobil Research and Development Corp., Paulsboro, NJ 08066.

0-8412-0383-0/78/33-166-083$05.00/0 © 1978 American Chemical Society

The model also predicted up to six stationary states in addition to the observed one. These seven states were in two groups: three in the upstream and four in the downstream region of the tube. Because of irreversible modification of the reactor by cementing in a liner with 4.76-mm i.d. to obtain laminar flow, the multiplicity predicted for turbulent flow could not be tested. They speculated that one of the additional predicted stationary states might exist physically but that the other five were probably mathematical artifacts of the idealizations in the model or in the method of solution.

Bernstein and Churchill (3) constructed an essentially identical combustion chamber, closely reproduced the experimental values of Chen and Churchill (1), and confirmed the existence of the predicted multiple stationary states. The exact number of stable stationary states remains somewhat uncertain, since the separation of the states within the two groupings was within the range which might be attributed to irreproducibility in setting the flow rates and inlet temperature. They also discovered that these flames produced only 5–32 ppm NO_x, with the particular value depending primarily on the location of the flame front inside the tube.

The primary objective of the work reported here was to determine whether flames from atomized fuel could be stabilized similarly in a refractory tube and whether such flames would also produce low concentrations of NO_x. To simplify analysis of the process, a single string of uniformly sized droplets was used. Two regimes can be conceived: one in which the droplets are completely evaporated prior to ignition and the other in which ignition occurs prior to complete evaporation. All of the results reported here fall in the first category.

Williams (4) has recently reviewed the extensive literature on the evaporation and combustion of fuel droplets. The environmental conditions considered here are quite different from those of all of these prior studies, and further citation does not appear to be appropriate. The development of a theoretical model for the process is in progress and will be reported subsequently.

Experimental Apparatus and Procedure

The experimental apparatus closely resembled that used by Chen and Churchill (1) and by Bernstein and Churchill (3) for premixed air and propane vapor, except that a longer combustion tube was used to assure complete evaporation and combustion. The apparatus consisted of a fuel supply and atomization system, an air supply system, and a combustion chamber with insulation and guard heaters as sketched in Figure 1.

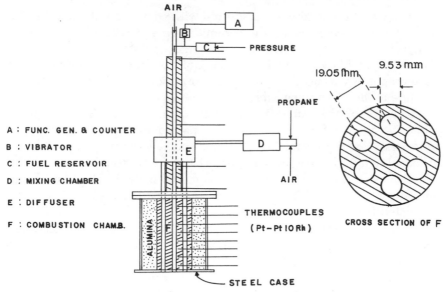

Figure 1. Schematic of experimental apparatus

The combustion chamber was a round channel 9.53 mm in diameter and 0.864 m long. The first 0.623 m was confined by a high purity, aluminum oxide tube. The final 0.241 m consisted of the central hole of an aluminum oxide block formed from Wulff furnace elements. The tube was insulated with 6.4 mm of powdered alumina surrounded by a larger aluminum oxide tube which was in turn insulated with glass fibers. Propane–air flames in six equidistant surrounding channels acted as guard heaters in the block. The block was insulated with powdered alumina.

Hexane was chosen as the fuel based on its vapor pressure. A more volatile liquid would require pressure storage or refrigeration; a less volatile fuel would require preheating the air or the use of a longer tube to accomplish evaporation before ignition. Reagent-grade hexane was pressured into the axis of the combustion channel through a 0.1143-mm (32 gage) stainless steel, hypodermic needle. The flow rate was measured with a float meter. Droplets 180–260 μm in diameter were generated by vibrating this capillary tube with the core of an audio speaker at a controlled frequency in the range of 6–9 kH.

Compressed air was filtered to remove oil and dirt, and the rate was measured with a Rotameter. The air entered the combustion tube concentric to the capillary tube.

The experiments were conducted at hexane-to-air mass ratios of 0.03–0.06, as compared with stoichiometric ratio of 0.066, and at total flow rates of 0.48–0.68 g/sec to assure turbulent flow upstream of the flame front.

The wall temperature was measured with copper/constantan thermocouples at four points along the ceramic tube and with platinum/platinum–10% rhodium thermocouples at eight points along the central channel in the block. The initial spacing of the droplets was determined photographically using a stroboscopic flash, prior to inserting the atomizing tube in the combustion tube.

Samples of the exit gas were withdrawn through 3.2-mm quartz tubing at about 0.8 cm³/sec. The gas passed through a condenser, a water trap, a tube containing Drierite, and a vacuum pump and was collected in a Teflon bag for analysis. CO and CO_2 were determined by non-dispersive infrared, O_2 by paramagnetic analysis, NO_x by chemiluminescence, and hydrocarbons by flame ionization. N_2 and H_2O were calculated from stoichiometry.

In order to establish a stable flame within the tube it was necessary first to preheat the ceramic block above the ignition temperature. This was done by burning gas from smaller tubes inserted in the seven channels of the ceramic block from the downstream end. The hexane–air flame attained a stationary position after about 7200 sec.

Experimental Results

Illustrative photographs of the string of droplets formed upon leaving the capillary tube are shown in Figure 2. The droplets attain a spherical shape in a short distance and are free of satellites.

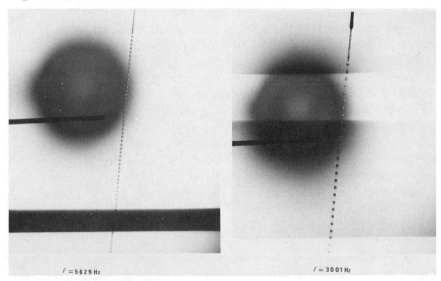

f = 5629 Hz   *f* = 3001 Hz

Figure 2. Formation of droplets. (Left). v_f = 0.01846 cm³/sec; f = 5.629 kH; D_d = 184 μm; $λ_d$ = 348 μm. (Right). v_f = 0.01846 cm³/sec; f = 3.001 kH; D_d = 227 μm; $λ_d$ = 622 μm.

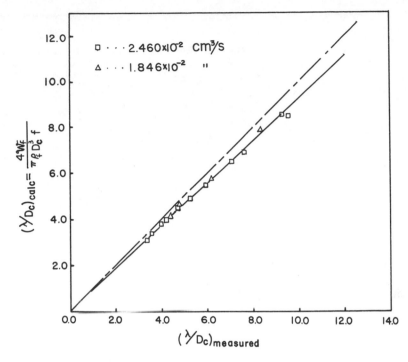

$$\left(\frac{y}{D_c}\right)_{calc} = \frac{4W_F}{\pi \rho D_c^3 f}$$

□ · · · 2.460×10² cm³/s
△ · · · 1.846×10⁻² "

$\left(\frac{y}{D_c}\right)_{measured}$

Figure 3. Test of droplet formation

The computed initial spacing of the droplets, assuming that the frequency of formation was equal to the frequency of vibration, is compared in Figure 3 with the values observed in photographs such as Figure 2. The observed spacing exceeds the computed spacing by a constant factor of about 1.07. This small discrepancy is probably caused by necking down of the liquid stream on leaving the capillary or by an error in one of the measured quantities. The spacing changes as the droplets are accelerated to the velocity of the gas stream. The diameter of the droplets was calculated simply from the measured rate of flow of hexane and the frequency of vibration, based on the above confirmation of one droplet per vibration.

Wall temperature profiles are plotted vs. distance down the combustion channel for a number of total flow rates and hexane-to-air ratios in Figure 4. The wall temperature decreases slightly at first, then rises slowly, then rapidly, and finally decreases slowly. As the flow rate is increased, the flame front, as reflected by the rapid rise in wall temperature, moves toward the exit of the tube, and the maximum temperature decreases. Decreasing the hexane-to-air ratio also shifts the flame front toward the exit.

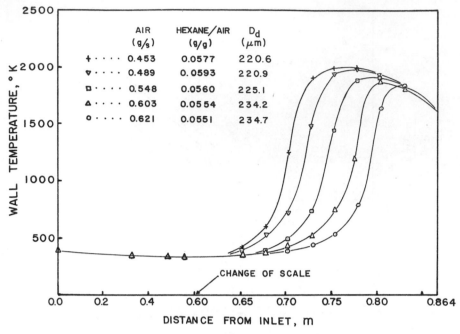

Figure 4. Wall temperature profiles

The measured values of NO_x and CO at the exit are plotted in Figure 5 vs. residence time at high temperature. These values of residence time were computed from the gas temperature profile down the tube, which was estimated from the wall temperature profile. The NO_x decreases, and the CO increases with decreasing residence time as might be expected from kinetic considerations. These very low values of NO_x and moderate values of CO are comparable in magnitude and trend with those of Bern-

Table I. Measured Gas Composition, Flame Front

Air Flow Rate (g/sec)	Hexane/Air (g/g)	Droplet Diameter (μm)	Flame Front Location (m from exit)
0.644	0.0569	235.8	0.107
0.602	0.0540	235.6	0.137
0.605	0.0559	243.4	0.114
0.634	0.0534	229.4	0.102
0.467	0.0552	223.2	0.168
0.621	0.0551	234.7	0.109
0.603	0.0554	234.2	0.135
0.489	0.0593	220.9	0.157
0.452	0.0577	220.6	0.175
0.456	0.0572	215.0	0.173

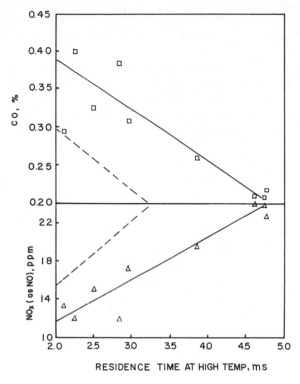

Figure 5. Exit gas composition. (- - -), Bernstein and Churchill (3) for premixed propane and air.

Location, and Residence Time at High Temperature

Residence Time (msec)	Exit Gas Composition	
	CO (%)	NO_x (as NO) (ppm)
2.07	0.325	11.3
2.95	0.305	17.15
2.41	0.325	14.8
2.09	0.295	13.1
4.61	0.211	24.0
2.25	0.400	11.8
2.83	0.385	11.8
3.90	0.230	19.5
4.77	0.210	22.7
4.70	0.214	21.7

stein and Churchill (3) for flames of premixed propane and air in the same combustion chamber, as indicated by the curves representing their measurements.

In addition to the above quantitative measurements, two qualitative observations appear to be relevant. The droplets evaporated completely prior to ignition of the air–vapor mixture, and the flames were quiet, stable, transparent, and essentially planar. The experimental conditions and observed values are summarized in Table I.

Discussion

A mathematical model for the process has not been completed yet. However the experimental results and the prior analysis of Chen and Churchill (1) for premixed propane and air suggest the following qualitative description as the fuel–air mixture passes through the combustion channel. The droplets are quickly accelerated to the velocity of the stream of gas and evaporate at a rate controlled by simultaneous heat and mass transfer with the surrounding gas. The gas temperature falls in the inlet region because of the convection of energy to the droplet. The wall temperature then falls because of convection of energy to the gas stream. Eventually the wall temperature begins to rise as heat transfer by conduction and radiation from the hotter downstream tube wall predominate over heat transfer by convection to the gas stream. The gas temperature then rises from convection from the wall; the rate of heat transfer from the gas to the droplet and the rate of evaporation increase correspondingly. When the droplet is completely evaporated, the wall and gas temperatures rise more rapidly. When the wall reaches the ignition temperature of the gaseous mixture, the gas temperature rises very rapidly because of combustion, exceeding the wall temperature and approaching the adiabatic flame temperature, then decreases by convection back to the wall. The maximum gas temperature may be slightly above or slightly below the adiabatic flame temperature depending on the relative magnitudes of thermal feedback, external heat losses, and incomplete release of the heat of combustion. The wall temperature rises, but more slowly than the gas, in the region of combustion, primarily because of convection from the gas stream. It reaches a maximum after the gas does and finally decreases because of radiant losses out the end of the tube.

When the flow rate is increased the flame front shifts downstream. The point of stabilization is controlled by a balance of the energy carried by the flowing stream and the other mechanisms of transfer mentioned above. The blowoff and flashback limits are attained when the energy lost from the combustor by radiation and carried out by the flow exceeds that released by the chemical reactions. The above description presumes that radial heat losses, longitudinal back-mixing, and radiant absorption and emission by the gas stream and droplets are negligible.

The droplets were evaporated prior to ignition of the vapor–air mixture in all of the work reported here. However operation may be possible with ignition prior to complete evaporation if the tube is sufficiently long. Complete evaporation of less volatile fuels prior to ignition may be possible by preheating the air and/or by using a longer combustion tube.

The close correspondence of the measured wall temperature profiles and exit-gas compositions to those of Bernstein and Churchill (3) for the combustion of premixed propane vapor and air suggests that combustion in a refractory tube is relatively insensitive to the composition and state of the fuel as long as evaporation precedes combustion.

The low NO_x production for combustion in a refractory tube apparently results from the short residence times of 2–5 msec at high temperature and from the minimal back-mixing caused by the planar flame front and the absence of acoustical oscillations. Bernstein and Churchill showed that the observed decrease in NO_x and increase in CO with decreased residence time are consistent with kinetic considerations.

Conclusions

Stable flames from atomized fuel droplets and air can be established inside a refractory tube over a range of flow rates, drop sizes, and fuel-to-air ratios. The wall temperature profiles and exit-gas compositions correspond closely to those for premixed fuel gas and air. Very low NO_x contents are attainable despite the high flame temperatures.

Further work is necessary to extend the range of flows to the blowoff and flashback limits for a series of droplet diameters, tube diameters, tube lengths, and fuel-to-air ratios. The possible existence of multiple stationary states for the same external conditions and the effects of nitrogen in the fuel, air preheating, ignition prior to complete evaporation, other fuels, and other atomization patterns should also be investigated.

Nomenclature

D_c = diameter of capillary tube, L
D_d = initial diameter of droplets, L
f = frequency of vibration, θ^{-1}
v_t = volumetric rate of flow of liquid hexane, L^3/θ
w_t = mass rate of flow of hexane, M/θ
λ_d = initial spacing of droplets, L
ρ_t = density of hexane, M/L^3

Acknowledgement

The analysis of the gas samples was carried out at the Mobil Research and Development Corp., Paulsboro, N.J. The authors are most grateful to V. W. Weekman and R. C. Murphy for this contribution. The authors are also appreciative of the partial support provided by NSF Grant Eng-75-20225.

Literature Cited

1. Chen, J. L.-P., Churchill, S. W., "Stabilization of Flames in Refractory Tubes," *Combust. Flame* (1972) **18**, 37–42.
2. Chen, J. L.-P., Churchill, S. W., "A Theoretical Model for Stable Combustion Inside a Refractory Tube," *Combust. Flame* (1972) **18**, 27–36.
3. Bernstein, M. H., Churchill, S. W., "Multiple Stationary States and NO_x Production for Turbulent Flames in Refractory Tubes," *Int. Combust. (Proc.) 16th* (1977).
4. Williams, A., "Combustion of Liquid Fuel: A Review," *Combust. Flame* (1973) **21**, 1–31.

RECEIVED December 29, 1976.

Potential Application of a Modulated Swirl Combustor to Clean Combustion of Liquid Fuel

A. K. GUPTA and J. M. BEÉR

Department of Chemical Engineering, Massachusetts Institute of Technology, Cambridge, MA 02139

A. C. STYLES

Department of Mechanical Engineering, University College, Cardiff, England

There are some fundamental differences between flame characteristics from a widely used industrial type of twin-fluid air-blast atomizer and from a modulated swirl combustor operated with a Sonicore nozzle atomizer. The twin-fluid atomizer shows significantly long dimensions of the flame in which combustion is completed by external entrainment into the flame from secondary air surrounding the flame. The modulated swirl combustor has at least two modes of operation designated as "ring" and "no-ring" which can be characterized by physical changes in flame characteristics. The combustor can be modulated to give either a bright yellow, highly radiative flame or a clean blue flame with the same air flow throughput. These spray flames burned as turbulent diffusion flames.

Turbulent flames are used widely for most domestic or industrial combustion processes. In the case of distillate fuels, e.g., liquid kerosene, detailed knowledge of droplet trajectories and rates of droplet burning plus statistical information concerning size and spatial distributions is necessary to describe completely the overall combustion processes, e.g., flame stabilization, carbon formation, mass burning rate, radiation properties, and pollutant emissions. Practical oil flames are evaporation and mixing-limited and may be considered as one of the two basic types:

(1) Turbulent jet diffusion flames. In this case, the oil is atomized by high-pressure air or steam (air blast atomizer), and the momentum of the fuel spray is so high that the entrained air is sufficient to complete the combustion.

(2) Pressure jet atomizer flames. In this flame, the momentum of the spray is small compared with the momentum of the air flow.
An understanding and analysis of single-droplet processes is necessary to understand oil combustion better.

Some results reported in this chapter were obtained using a twin-fluid air-blast atomizer (e.g., distribution of mean temperatures, degree of combustion) and showed significantly long dimensions of the flame at all flow throughputs and mixture ratios. Combustion is completed by external entrainment to the flame from secondary air surrounding the flame. The average drop size of the liquid fuel using this atomizer was large, typically of the order of 110 μm. A fine, fog-like spray can be obtained using an ultrasonic whistle nozzle. This nozzle focuses high-frequency sound waves which produce an oscillating pressure wave. This oscillating wave in the resonant chamber of the nozzle atomizes the liquid injected into the cavity of the resonant chamber.

Gas turbine combustors, normally operated with kerosene fuel, are required to operate with high efficiency and low pollution over a wide range of conditions. To achieve low CO and NO_x emissions requires that the primary zone operate at an overall equivalence ratio of about 0.6–0.8 of stoichiometric (i.e., weak). Although this condition can be met at any one design operating point, such as the take-off power setting, the mixture ratio would then be too weak at idle with a simple fixed-geometry combustor. There are several techniques known to control emission of NO_x, such as two or multi-stage combustion (1) or dilution of the reaction zone by recirculation. Reduction in NO_x emission levels from high nitrogen content fuels can be obtained either by using a staged combustion system or by reducing the conversion of fuel nitrogen to NO_x by increasing the combustion temperatures. We, therefore, require staged fuel and/or air addition to the combustor to attain low pollution and high efficiency throughout the operating range.

Potential mechanical problems associated with mechanically variable geometry combustors have promoted studies on various types of aero-dynamically modulated combustors. This modulation of the combustor could, in principle, be controlled by very small control flows introduced or extracted from the chamber of the combustor. Alternatively, it could be inherent in the aerodynamics of the combustor. Flexibility in combustor operation can be obtained easily by small changes in combustor geometry so as to obtain the desired aerodynamic changes within the combustor. It can also be obtained by changing the stoichiometry of the flow.

The modulated swirl combustor decribed here (2) represents a step in this direction when operated with propane or liquid kerosene. It exhibits a very distinct change in flow pattern and potentially can satisfy pollution legislation over a wide operating range of heat release rates, as required, for example, in an aircraft gas turbine engine combustor.

Swirl Combustor for Flame Stabilization

Swirl is used commonly to stabilize high-intensity combustion (3, 4, 5, 6, 7). In general, there are two main types of swirl combustors.

(1) The swirl burner. Here the swirling flow exhausts into the atmosphere or a cavity (e.g., a combustion chamber). The combustion in swirl burners occurs mostly in and just outside the burner exit.

(2) The cyclone combustion chamber. Here the air is injected tangentially into a usually cylindrical chamber and exhausts through a centrally located exit hole at one end. The cyclone combustors are usually used to burn efficiently difficult-burning materials such as high-ash-content and brown coals, anthracite, high-sulfur-content oils, damp vegetable refuse, poor-quality-low-calorific-value fuels, etc.
Very high residence times are achieved with cyclone combustors, and most of the combustion process occurs within the chamber.

Swirling flows confined in cylindrical chambers exhibit a variety of different flow patterns. In some cases a small control flow can cause a significant change from one pattern to another. An extreme form of this was shown in films made at the Jet Propulsion Laboratories in California in which longitudinal secondary flows in a cylinder were altered abruptly by inserting very fine wires across the chamber.

Experimental Apparatus

Twin-Fluid Air-Blast Atomizer. Twin-fluid atomizers can be divided into internal and external mixing systems. Atomization occurs by passing a high-velocity gas stream over a liquid sheet or by mixing in the form of a "Y" jet. The gas stream is usually air although steam has been used to improve the injection characteristics of heavy viscous fuels. The air stream is usually derived from the main air flow to the combustor, thus utilizing a portion of the combustor pressure drop.

The twin-fluid atomizer used here and described elsewhere (8) is based on the above principle in which liquid fuel and air are forced through a nozzle and emerge from the orifice in the form of a spray cone. A schematic of the atomizer is shown in Figure 1.

Modulated Swirl Combustor. The design of the modulated swirl combustor evolved from swirl burners tested by the authors at Sheffield University and Institute of Flame Research Foundation (IFRF), Ijmuiden (3, 4, 5, 9, 10). The modulated swirl combustor (Figure 2) consists essentially of a cylindrical chamber 1.8 diameters long with an outlet consisting of a contraction followed by a short wide-angle diffuser. Air could be

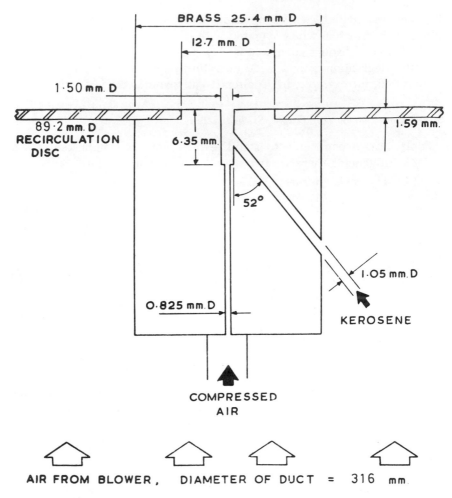

Figure 1.   Twin-fluid atomizer.   The kerosene–air system.

introduced tangentially into the chamber at four equally spaced stations, i.e., 1, 2, 3, and 4. At station 4, the introduction of air into the chamber was at 20° to the tangent (Figure 2). Flow at each pair of inlets into the combustor could be adjusted independently. The facility introduced gaseous fuel (propane) into the combustor either at the base or tangentially at station 3. The gaseous fuel injection at the base of the combustor could be replaced by liquid fuel in which fine, fog-like droplets were produced by an ultrasonic atomizing nozzle. Ultrasonic whistle atomization is a relatively new technique for breaking up viscous liquids. Here instead of forcing a liquid through a small orifice under high pressure, the Sonicore atomizer accomplishes breakup in an intense field of sonic energy. Any liquid delivered to the sonic energy can be atomized into a very fine, fog-like spray. A sketch of the ultrasonic whistle nozzle atomizer is shown in Figure 3.

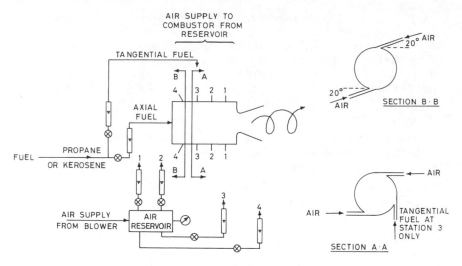

Figure 2. Schematic of the modulated swirl combustor

Details of the modulated swirl combustor are as follows:

Length of the chamber $= 114.3$ mm
Diameter of the chamber $= 63.5$ mm
Diameter of the throat $= 38.1$ mm

$\dfrac{d_{\text{throat}}}{D_{\text{exit}}} = 0.6$

Angle of divergence
of the nozzle $= 40°$.

The results reported in this chapter are confined to two fuels: propane gas supplied tangentially at station 3 and liquid kerosene supplied with an ultrasonic whistle nozzle atomizer on the central axis at the base of the combustor.

The multiplicity of individually controlled air inlets is, of course, an experimental expedient. In a fully developed combustor each air inlet could be replaced by a correctly sized hole, or set of holes, to give the same volume and momentum flux as in the experimental combustor.

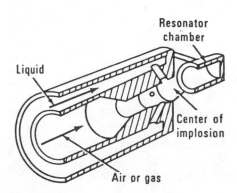

Figure 3. The Sonicore nozzle atomizer

The combustion stability limits, size of the recirculation zone, completeness of combustion, and the radial distribution of mean temperatures have been measured. Mean temperatures were measured with a 0.076-mm Pt vs. Pt–13% Rh thermocouple coated with silica to reduce catalytic effects. The completeness of combustion was measured with gas chromatography using a Pye Unicam isothermal gas chromatograph. A detailed description of the measurement technique used is given in Refs. 5 and 9.

Theoretical Characterization of Spray Combustion

In most practical systems, e.g., combustors or furnaces, the combustion process is much more complicated because of two important factors. First, to a large extent the fuel and oxidant are mixed within the quarl of a burner or within the combustion chamber itself. The mechanism of the reactant mixing therefore plays an important role. Second, the flow aerodynamics are complicated by the flow recirculation and turbulence and frequently cannot be represented by simplified models.

However, despite the complexity of this scheme much valuable information has been obtained by applying some simplified analysis. In all these cases a major problem that has been considered, and to a large extent solved, is concerned with the combustion mechanism of individual droplets that make up the spray. In particular it has been necessary to study the nature of the combustion process and whether combustion occurs in the gas phase or on the surface of the droplet and whether the rate is controlled by vaporization or by the kinetics of the chemical reactions involved.

In order to understand fully the processes involved in spray combustion, it is necessary to have a complete knowledge of:

(1) The mechanism of combustion of the individual fuel droplets,

(2) Droplet interaction under combustion conditions,

(3) Statistical distribution of the droplets, i.e., size and spatial distribution, and

(4) The flow aerodynamics.

In the aerodynamic effects, a knowledge of the drag coefficients of the individual droplets that make up a spray is essential to calculate the penetration of a spray into a combustion chamber and the subsequent movement of the burning droplets. Generally, however, information is necessary on evaporating rather than burning droplets because usually droplet velocities are greatest in the injection region where evaporation, rather than combustion, is the dominant factor. Burning and evaporating droplets differ from that ideal model in that when vapor effuses from the droplet surface the skin, drag decreases because of the thickened boundary layer. The relative effects are influenced by the type of flame-wake or envelope and by whether the droplet is just vaporizing. As a consequence of these effects the drag coefficient of an evaporating or burning droplet is less than the standard curve for an ideal sphere.

The most commonly considered case of droplet combustion involves the combustion of a liquid fuel burning in a surrounding oxidizing atmosphere, usually air. The droplet evaporates and acts as a source of vapor, and, since oxidant and fuel are initially separated, the fuel vapor and oxidant burn in a diffusion flame surrounding the droplet.

Recently, Mizutani et al. (*11*) presented data on an analogous spray diffusion flame, stabilized in a heated air stream. Pressure-atomized injection was used with a moderate swirl in the fluid stream to obtain a full cone spray. The two-phase flow was sampled, and the liquid fraction was determined by separation and weighing. The spray, having some hollow cone characteristics, had maximum droplet flux at some distance from the axis of the flow. The cool core region had relatively low drop evaporation rates and did not indicate strong combustion effects. The fuel vaporized in this region was transported to the flame zone by turbulent mixing, similar to that in a gaseous diffusion flame.

Onuma (*12*) studied the structure of a spray combustion flame. The results are qualitatively similar to the findings of Mizutani (*11*). Onuma, furthermore, showed that a good similarity exists between the spray combustion flame and a turbulent gas diffusion flame.

It is therefore evident that a detailed experimental analysis of the spray flame is necessary for its theoretical characterization. The aerodynamic changes achieved within the modulated swirl combustor, in which it is demonstrated that blue flame combustion of oil can be achieved using this combustor and a Sonicore atomizing nozzle, clearly show the strong effect of flow aerodynamics upon the spray combustion process.

Experimental Results and Discussions

Twin-Fluid Atomizer. The typical distributions of mean temperatures and species concentrations from a twin-fluid atomizer are reported in this section. Figure 4 depicts the radial distribution of mean temperatures at various axial distances away from the spray nozzle exit having a fuel flow rate of 2 kg/hr and an atomizer air/fuel ratio of 0.32. It can be seen from Figure 4 that the shape of the temperature profiles remains the same at axial distances up to 200 mm downstream of the exit having a sharp peak at $r = 12$ mm. This shows that the burning occurs essentially just outside and near to the spray boundary which was later confirmed using first a still photography and then a high-speed cine photography, in agreement with the findings of Mizutani (*11*). No evidence was obtained of the individual droplets burning, Figure 5. In the intial regions of the flame there is a clear evidence that the droplet evaporates and that the flame surrounds these vapors and burns like a turbulent diffusion flame. Lack of oxygen on the central axis (confirmed by gas analysis) leads to uncracked gaseous hydrocarbons and large concentrations of soot which

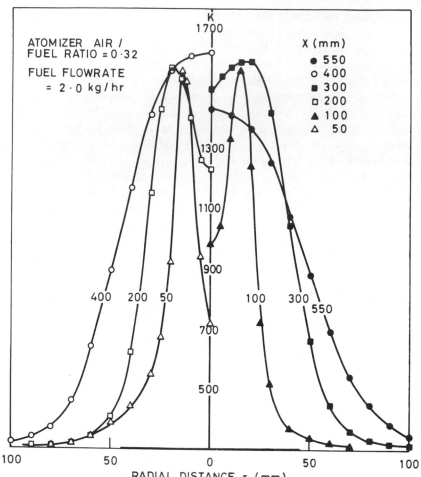

Figure 4.   Radial distribution of mean temperatures at various distances
along the kerosene spray flame—twin-fluid atomizer

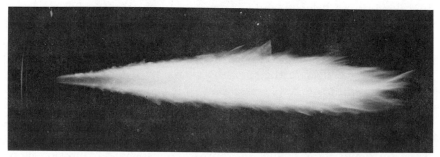

Figure 5.   Flame photograph—
twin-fluid atomizer kerosene spray
flame

are at a lower temperature than the main reaction zone. However, with increasing axial distance and the spread of the spray, temperatures on the central axis increase as the heat from the spray boundary is carried to the center by the entrained air. Maximum mean temperatures were found on the central axis of the spray flame at an axial distance of 400 mm downstream of the nozzle exit. Figure 6 shows the radial distribution of species concentration at $X = 500$ mm downstream of the nozzle, indicating the quality of combustion obtained with this type of atomizer. The spray flame is analogous to a gaseous diffusion flame where the main

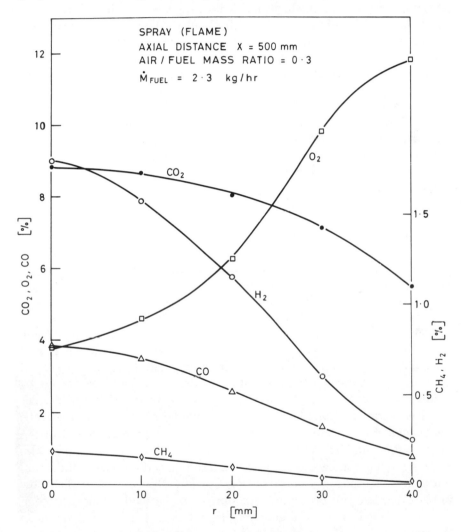

Figure 6.   Radical distribution of species concentration—twin-fluid atomizer kerosene spray flame

reaction zone, and therefore maximum mean temperature, is displaced radially from the central axis of the jet, Figure 4. Further downstream where the flame is no longer a heterogeneous mixture (comprising vaporizing droplets and combustion gases) and burning is only intermittent, intermediates of combustion, e.g., CH_4, CO, and H_2, are still measurable. Carbon dioxide concentrations are high in the center of the flame with oxygen concentrations remaining low owing to reaction with the intermediates to form CO_2 and H_2O.

Modulated Swirl Combustor Characteristics. Many preliminary tests were carried out in which the air flow rates were adjusted in various combinations for different modes of fuel entry. These tests showed that the most significant flow pattern changes occurred with the tangential fuel entry, and, furthermore, the combustor had two distinct modes of operation. Under certain conditions of air distribution and loading, the combustion zone was confined to a layer close to the walls of the chamber and extended out into a closed recirculation bubble in the outlet nozzle, Figure 7, designated as "no-ring." The flow form in this type of flame is similar to that in a conventional swirl combustor as reported in Refs. 3, 4, and 5. Very little combustion takes place within the combustor; most of it occurs within the nozzle and past the exit of the combustor. A slight change in either the air distribution or loading alters the shape of the flame producing a combustion zone in the form of a distinct cylindrical shell. This cylindrical shell lies near to the central axis away from the walls of the nozzle and has a diameter about half that of the chamber designated as "ring" (Figure 7). The size of the recirculation zone downstream of the diffuser was reduced significantly, and noise emission increased in this second mode of operation.

As far as could be determined, the two modes of combustion, viz. no-ring and ring, depended entirely upon the air and fuel distribution, implying that there was no hysteresis in the characteristics. The main objective, however, was to determine the steady-state performance of the combustor. Details of the switch from one state to another were not investigated. A change from one state to the other could be caused by holding the fuel flow constant and changing the air distribution; the pair of inlets at stations 1 or 4 having the most significant effect. Alternatively, the change of flow state could be caused by holding the air flow distribution constant and varying the fuel flow. The ring and no-ring modes of burner operation have been observed also in a certain type of industrial burners (13) used in cement industry for low NO_x emission. The flow in these burners can be modulated, similar to the modulated swirl combustor presented here. The geometry is, however, somewhat different in that oil spray and air are introduced into primary and secondary regimes. The primary regime consists of a central spray oil gun surrounded by

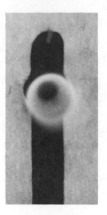

Figure 7. Flame photographs obtained from the modulated swirl combustor using propane as the fuel showing no-ring and ring modes of combustion

primary swirling flow. The secondary flow is introduced surrounding the primary flow via six small swirlers having separate controls for air and fuel flow.

The different type of combustion processes obtained from this experimental modulated swirl combustor therefore help in understanding and analyzing the complex flow behavior observed in certain type of industrial flames (*13*).

PROPANE FUEL—RESULTS AND DISCUSSION. Onuma (*12*) showed that in a kerosene spray flame, there is no evidence of droplet burning. The vapor cloud formed by evaporation of the droplets burns like a turbulent diffusion flame. A close relationship between kerosene spray flame and gaseous diffusion flames (using propane as the fuel) was provided. The results reported in this section are those obtained from the modulated swirl combustor using propane as the fuel.

Tests were carried out over a range of air flow rates and mixture ratios. For a particular fixed air distribution, the stability limits and the ranges of the two modes of combustion are shown in Figure 8. At any volume of air flow, as the fuel flow is increased beyond the weak blow-off limit, the first no-ring mode occurs, then there is an abrupt change to the ring mode followed eventually by another abrupt change to the no ring mode. The equivalence ratio given by:

$$\phi \, \frac{(\text{Fuel/air}) \text{ actual}}{(\text{Fuel/air}) \text{ stoichiometric}}$$

was calculated when an abrupt change of flow state was observed. From Figure 8 it is apparent that the two modes are determined largely by the stoichiometry of the system. A rich "blow-off" limit could not be determined. However, in this combustor, this was not significant since stable burning took place well outside what would normally be the confines of the combustor.

The size of the recirculation zone downstream of the exit nozzle for the two modes of combustion was determined by introducing salt solutions through a small hypodermic tube and is shown in Figure 9. The total air flow through the modulated swirl combustor was held at 1070 L/min, and the fuel flow was set at 30 L/min for no-ring and at 55 L/min for ring. The larger recirculation zone, hence larger recirculating mass flow, occurs with the no-ring mode of combustion.

Radial distribution of mean temperatures were measured in the main

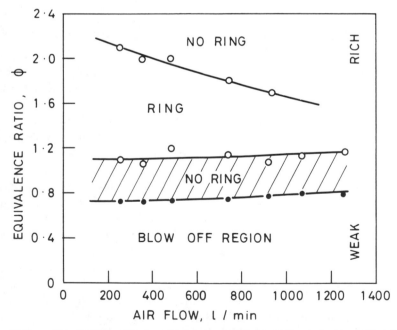

Figure 8. Stability limits of the modulated swirl combustor. Total air flow = 1070 L/min, equally distributed in the four points of inlets.

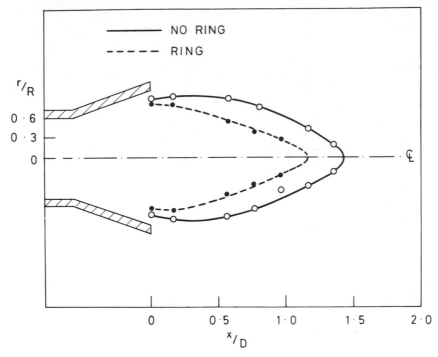

Figure 9. Boundaries of the internal recirculation region in the modulated swirl combustor—propane fuel

combustion region just downstream of the nozzle exit for the two modes of combustion. The results are depicted in Figure 10. The temperature distributions were symmetrical about the central axis and reflect the two different forms of recirculation zone, i.e., relatively less entrainment for the case of the ring mode gives much higher mean temperatures as expected. With the ring mode of combustion, the mean temperature profiles are much fatter than with the no-ring mode of combustion. The temperature of the chamber wall at all operating conditions was less than about 700°C showing an effective film cooling by the swirling air which remains attached to the walls of the combustor because of centrifugal force.

Measurements of the species concentration (CO_2, H_2, O_2, and CO) with and without a ring at the nozzle exit and at two diameters downstream of the exit are shown in Figure 11. The main features of the measurements are consistent with other results (e.g., size of the recirculation zone and mean temperatures), and the profiles are similar to those for conventional swirl-stabilized flames. The central recirculation zone consists largely of hot burnt products (i.e., CO_2). There are, however, significant differences in the two modes, e.g., in the ring mode, com-

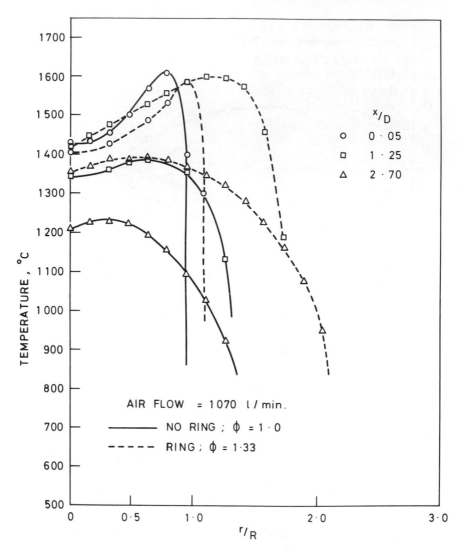

Figure 10. Radial distribution of mean temperatures in the no-ring and ring modes of combustion—propane fuel

bustion is completed in a shorter length (compare CO concentrations level at $x/D = 2.0$ for the two modes of operation).

KEROSENE FUEL—RESULTS AND DISCUSSION. Liquid kerosene was burned in the above modulated swirl combustor using a Sonicore atomizing nozzle. As expected the flame could be modulated to obtain a bright yellow, highly radiative flame or a clean blue flame as shown in Figure 12. Total air flow was the same, and the burn-out was complete by about 1 diameter downstream of the burner exit in both cases. Radial distribution of mean temperature at various axial distances within and just outside the combustor for the two cases (bright yellow and blue flame) is shown in Figure 13. The profiles are similar to those obtained with the gaseous propane fuel. The presence of a relatively large internal recirculation zone for the case of blue flame combustion conditions can be easily recognized by the presence of lower mean temperatures within the flame. The relative size of the internal recirculation zones for the two cases of bright yellow and blue flames were investigated and were found to correspond to the ring and no-ring mode of combustor operation.

Conclusions

The size, shape, and length of the flames obtained from the two types of atomizers (twin-fluid air blast and Sonicore) show some fundamental differences between the two atomizers. A relatively large droplet size from the twin-fluid atomizer together with low concentrations of oxygen and temperature within the spray allow combustion to occur essentially at the spray boundary.

The modulated swirl combustor presented here has generally stable and acceptable basic characteristics, but in addition, it has two modes of

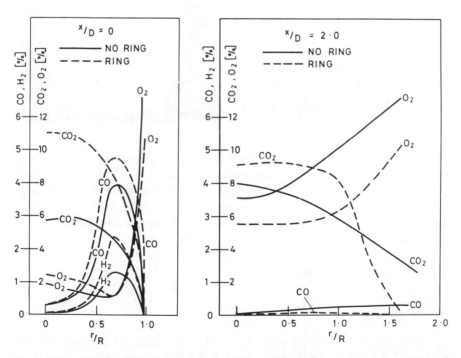

Figure 11.   Radial distribution of species concentration in the no-ring and ring modes of combustion—propane fuel

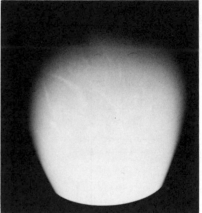

Figure 12.   Flame photographs of the kersosene–air flame using "sonicore" atomizer and the modulated swirl combustor. (a) Bright yellow, highly radiative flame; (b) clean, blue flame.

operation. These modes are separated by slight changes in air distribution or fuel flow and imply a large modulation as a result of small control signal. The two modes of combustion can be characterized by a physical change in the flame shape, temperature distribution, or composition distribution, and by a change in the noise emission levels. Tests carried out using liquid kerosene fuel and a Sonicore nozzle atomizer show that the combustor could be modulated to give either a bright yellow, highly radiative flame or a clean blue flame.

Acknowledgments

The authors acknowledge gratefully the financial support of the Science Research Council and the A.E.R.E. Harwell. Discussions and advice of D. S. Taylor, J. R. Tippetts, and N. Syred are gratefully ack-

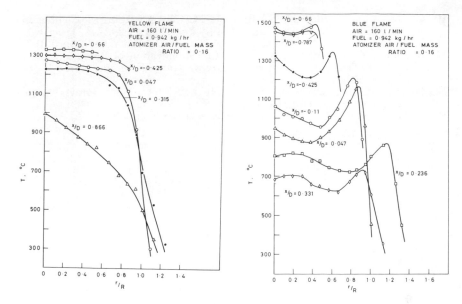

Figure 13. Radial distribution of mean temperatures corresponding to flames as in Figure 12

nowledged. Thanks are also due R. Mughal. The work was carried out at Sheffield University, Department of Chemical Engineering and Fuel Technology, England.

Literature Cited

1. Gupta, A. K., Syred, N., Beér, J. M., "Fluctuating Temperature and Pressure Effects on the Noise Output of Swirl Burners," *Symp. (Int.) Combust. (Proc.) 15th* (1975) 1367–1377.
2. Gupta, A. K., Tippetts, J. R., Swithenbank, J., "Modulated Swirl Combustor," 2nd European Symposium on Combustion, Orleans, pp. 690–696, September 1975.
3. Beér, J. M., Chigier, N. A., "Combustion Aerodynamics," Applied Science Publishers, London, 1974.
4. Syred, N., Beér, J. M., "Swirl Stabilized Combustion—a Review," *Combust. Flame* (1974) **23** (2) 143–202.
5. Gupta, A. K., Lilley, D. G., Syred, N., "Swirl Flows," Applied Science Publishers, London, 1978.
6. Mather, M. L., MacCallum, N. R. L., *J. Inst. Fuel* (1967) **40**, (316) 214–245.
7. Beltagui, S. A., MacCallum, N. R. L., "Vane-Swirled Flames in Furnaces," 2nd European Symposium on Combustion, Orleans, pp. 672–677, September 1975.
8. Roett, F., Ph.D. Thesis, Department of Chemical Engineering and Fuel Technology, Sheffield University, 1972.
9. Gupta, A. K., Ph.D. Thesis, Department of Chemical Engineering and Fuel Technology, Sheffield University, 1973.

10. Gupta, A. K., Taylor, D. S., Beér, J. M., "Investigation of Combustion Instabilities in Swirling Flames Using Real Time L.D.V.," "Proc. Symposium on Turbulent Shear Flows," Penn State University, April 1977.
11. Mizutani, Y., Yasuma, G., Katsuki, M., "Stabilization of Spray Flames in a High Temperature Stream," *Symp. (Int.) Combust. (Proc.) 16th* (1977), 631–638.
12. Onuma, Y., Ogasawara, M., "Studies on the Structure of a Spray Combustion Flame, *Symp. (Int.) Combust. 15th*, Tokyo (1975) 453–465.
13. Gupta, A. K., N. F. K. Ltd., Japan, private communication, 1976.

RECEIVED December 29, 1976.

Temperature, Concentration, and Velocity Measurements in Fuel Spray Free Flames

NORMAN A. CHIGIER and A. C. STYLES

Department of Chemical Engineering and Fuel Technology,
University of Sheffield, England

*Measurements have been made of the combustion charac-
teristics of an air blast kerosene spray flame and of droplet
sizes within the spray boundary of isothermal sprays. Specific techniques were used to measure velocity, temperature,
concentration, and droplet size. Velocities measured by
laser anemometer in spray flames in some areas are 400%
higher than those in isothermal sprays. Temperature profiles are similar to those of gaseous diffusion flames. Gas
analyses indicate the formation of intermediate reactants,
e.g., CO and H_2, in the cracking process. Rosin–Rammler
mean size and size distribution of droplets in isothermal
sprays are related to atomizer efficiency and subsequent
secondary atomizer/vaporization effects.*

Combustion of liquid fuels in combustion chambers follows the atomization of bulk liquid into droplets, interaction of the spray of droplets with atomizing and combustion air streams, vaporization, cracking
and formation of intermediate products, combustion in the vapor phase,
and, finally, dilution of combustion products and unburned reactants
with excess air. The vaporization rates of the lighter and more volatile
fractions of fuel mixture depend on the vapor/liquid equilibria and mass
transfer conditions within the droplet. The heavier fractions, particularly
those with asphaltene content, can persist in droplet or particle form for
long periods and can burn as single liquid or solid particles. The characterization of a spray flame is based on the changes in velocity, temperature, species concentration, liquid, and particle size as they vary from the
initial conditions at the atomizer exit to the side and end boundaries of
the spray and flame.

0-8412-0383-0/78/33-166-111$05.00/0 © 1978 American Chemical Society

This chapter reports on a set of measurements and measurement techniques made in air-blast kerosene spray flames fired vertically upwards in an unconfined air space. A laser diffraction meter was used to measure the variation of size distribution of droplets along the length of the spray as a function of variation of air and fuel flow rates. A laser anemometer, using a rotating-disk diffraction grating for frequency shifting and a digital pulse counter interfaced with a computer for data acquisition and analysis, was used for velocity and turbulence measurements in the spray flame. Liquid droplets and soot particles provided adequate seeding for laser anemometer measurements within the spray, but magnesium oxide particles were added to the secondary air to measure the velocity of the air flow. Comparisons of temperature measurements in the spray flame by suction pyrometer and coated thermocouples demonstrated the higher accuracy of the thermocouple in the range 1100–1600 K. Profiles of measured concentrations of CH_4, H_2, O_2, CO, and CO_2 show the extent of vaporization, cracking, and formation of intermediate combustion products, degrees of oxidation, and mixing rates of reactants, intermediates, combustion products, and air. All measurements provide time-average quantities, and the relationship between the time-average quantities and time-dependent variations of velocity, temperature, species concentration, and rates of evaporation and burning of individual droplets requires further study with high-frequency response instrumentation.

Atomizer and Air Flow System

An air-blast atomizer, designed on the basis of industrial-type twin-fluid atomizers, was supplied with kerosene under low pressure from a nitrogen cylinder and under high pressure air from a compressor. The atomizer nozzle diameter was 1.5 mm, a stabilizer disk with 89-mm diameter was fitted into an annular air duct of 316-mm diameter, kerosene flow rate was 2.3 kg/hr, and atomizing air flow rate was 0.69 kg/hr.

Air from the compressor enters the mixing chamber of the atomizer at sonic velocity and, after interaction with the liquid kerosene stream, emerges as a two-phase mixture, directed vertically upwards. The air flow from the annular stream forms a recirculation zone in the wake of the stabilizer disk. The flame is ignited by an external gas stream and subsequently burns independently as a flame in the open atmosphere. Droplets are initially confined to the air jet from the atomizer nozzle, but some of the finer droplets are taken up by the reverse flow of the stabilizer disk recirculation zone. Previous studies on spray combustion and details of atomizer design are reviewed by Chigier (1).

Velocity Measurements

Previous attempts to measure velocity in spray flames have been made by water-cooled pitot tubes. However the orifices of these tubes are often blocked by liquid and solid particles. Hetsroni (2) has succeeded in making velocity measurements of droplets in nonburning sprays by using a hot wire anemometer, but this method cannot be used in dense sprays containing large droplets and is not suitable for measurements in flames. High-speed photomicrography, using high-energy double-flash systems, has been successfully used to measure velocity, direction, and size of individual droplets within sprays and spray flames (3). Analysis of photographs of droplets and particles in flames has been facilitated greatly by use of the Quantimet electronic scanning pattern recognition instrument, which allows rapid and automatic analysis, providing print-out of the sizes of all droplets on the photograph.

Rapid developments have taken place in the field of laser anemometry, and this technique has been applied successfully in a number of studies on measurements in gaseous flames. In these studies, the gas flow was seeded with micron or submicron particles, and the velocity of these particles was taken to be representative of the velocity of the local gas flow. For the study reported here, a laser anemometer was adapted for the special problem of measurements in a spray flame which initially contains a polydisperse cloud of droplets up to 300 μm in diameter. Droplets and carbon particles are present, and seeded particles are added to the annular air flow. For the particles larger than 1 μm, significant differences exist between velocities of particles and surrounding gas. A complete description of the velocity field requires simultaneous measurement of velocity and size of individual particles. This has not yet been achieved, and, for this study, the velocity of all particles passing through the measurement control volume of the laser anemometer are reported.

The laser anemometer, as used for measurements in the spray flame, is shown in Figure 1. The anemometer is made up of a 500-mW argon ion laser, transmission and receiving optics, and a digital signal processing system, interfaced to a PDP-8E computer. Laser light, with 488-nm wavelength, is split by a bleached diffraction grating, and the first-order light beams are focused to a measuring control volume of 0.3 $\times$ 0.3 $\times$ 2 mm. Interference fringes are formed in the volume of the two crossed beams, and the beat frequency of light scattered by particles passing through the light and dark regions of the fringe pattern is measured. The frequency is directly proportional to the velocity component of the particle, normal to the anemometer optical axis, and in the plane of the two beams:

$$u = \Delta x v \qquad\qquad (1)$$

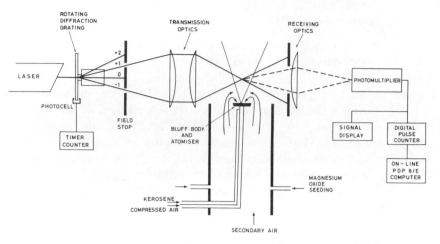

Figure 1. Laser anemometer for measuring velocity in an air blast spray flame

where u = velocity component of particle, ν = measured frequency, and Δx = fringe spacing. For digital signal processing, the frequency measured by the photomultiplier is determined by:

$$\nu = \frac{\text{number of fringes crossed}}{\text{time taken to cross fringes}} \qquad (2)$$

The annular air flow was seeded with magnesium oxide particles with a nominal diameter of 3–5 μm. Significant droplets and particles were available in the main spray to obtain good signal-to-noise ratios.

Light scattered from the droplets or particles passing through the control volume is collected by the receiving optics and focused onto the cathode of a photomultiplier. The frequency modulated output is fed into a digital pulse counter and processed as in Equation 2 to determine the instantaneous velocity of individual particles. With an interfaced PDP-8E computer, velocities of individual particles are stored as individual realizations or as a group sample in the form of a probability density distribution. This facility has greatly improved the data acquisition storage and sampling rates since that reported in Ref. 4. Signal processing is now capable of sampling rates of 16 kHz. However, for the present work, data acquisition was limited to some 2000 readings at each measuring point. Results were analyzed by the computer to determine the probability density distribution, mean velocity, and standard deviation.

In order to make measurements in reverse flow regions and in regions where relative turbulence intensities exceed 30%, a frequency shifting device was used. The frequency shift is determined by the rotation speed of the radial diffraction grating and by the order of diffraction of the beam, i.e.:

$$\nu_s = 2fN_G\,m \qquad\qquad (3)$$

where ν_s = frequency shift, f = rotational frequency of disk, N_G = number of radial lines on grating, and m = diffraction order of beam. The

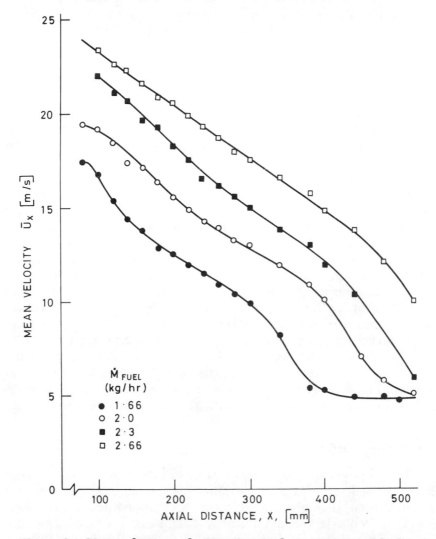

Figure 2.   Mean velocity on the axis of spray flame as measured by laser anemometer. Fuel and air flow rates varied with the ratio constant at 0.3.

flow velocities are determined by subtracting the shifted frequency ν_s from the measured frequency ν i.e.:

$$u = \Delta x (\nu - \nu_s) \qquad (4)$$

when $\nu_s > \nu$, then $u < 0$. Sufficient frequency shift was imposed on all velocities so that all negative velocities were measured as positive occurrences.

The axial component of mean velocities $\overline{U}_x$ on the centerline of four spray flames is shown in Figure 2. Air and fuel flow rates were altered while the air/fuel ratio at the atomizer exit was maintained constant at 0.3. As the fuel and air flow rates were increased, corresponding increases in mean velocity were detected along the length of the spray. The presence of liquid droplets and the effects of temperature increase both result in a rate of velocity decrease on the axis which is less than that of a corresponding air jet.

Isovelocity lines are plotted for both the isothermal spray and the spray flame in Figure 3. The cloud of droplets in the initial central region of the spray was too dense to allow velocity measurements. The flow field is comprised of a central air jet and spray with an annular toroidal recirculation zone in the wake of the stabilizer disk. Negative velocity in the reverse flow zones could be measured with the air of the frequency shift in the laser anemometer. Comparison of the flow fields in the spray flame with that in the isothermal spray (Figure 3) shows substantial increase in velocity magnitudes as a direct consequence of combustion. These velocity increases arise from increases in volumetric flow rate resulting from temperature increases as well as from the increased volumetric flow rate of fuel vapor as compared with liquid fuel. The acceleration of flame gases can be seen by comparing the velocity at $X = 400$ mm on the axis of the spray flame (14 m/sec) with that for the nonburning spray (3 m/sec). Evidence also can be seen of lateral expansion of the spray flame as a consequence of increase in volumetric flow rate. Combustion also causes changes in the recirculation zone, where maximum reverse flows are increased, as well as in the size of the recirculation zone.

Temperature Measurements

Special problems arise in measuring local temperature within spray flames. Liquid and solid particles cause deposits and blockage of orifices in instruments. High-temperature conditions, with particles having high emissivity, result in complex radiative heat transfer which affects the accuracy of temperature measurement. In industrial furnaces and gas turbine combustion chambers, suction pyrometers have been used for local temperature measurements. In the suction pyrometers, the thermo-couple is covered with a ceramic sheath to protect against mechanical damage and chemical interaction with flame gases. In addition, one or more shields surround the sheath to reduce radiative transfer while gases are sucked at high velocity from the flame past the sheathed thermo-couple. Water cooling also generally must be supplied to part of the probe. Despite the attempts to reduce the size of the components, suction pyrometers disturb considerably flow and temperature fields within flames.

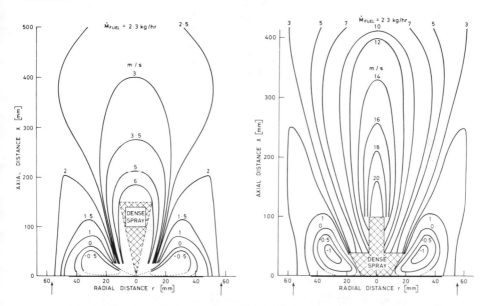

Figure 3. (a) (left) Isothermal spray; (b) (right) spray flame. Isovelocity lines measured by laser anemometer in isothermal spray and spray flame.

For this study, a comparison has been made between measurements with a miniature suction pyrometer and with thermocouples coated with silica. A platinum/13% platinum rhodium thermocouple with 75-μm diameter wires and 350-μm support wires was threaded through a twin bore refractory tube. The junction of the fine wires was flame-welded to give a bead size of approximately 100 μm. The thermocouple bead was coated with approximately 25 μm of silica. Comparison of suction py-rometer and thermocouple measurements made at axial stations $X = 100$ and 200 mm are shown in Figure 4. Temperature measurements below 1000 K show no significant difference when measured by the suction pyrometer or thermocouple. In the temperature range 1000–1600 K, the thermocouple always measured higher temperatures, up to 200 K. In the high-temperature regions, there are high levels of particle and soot con-

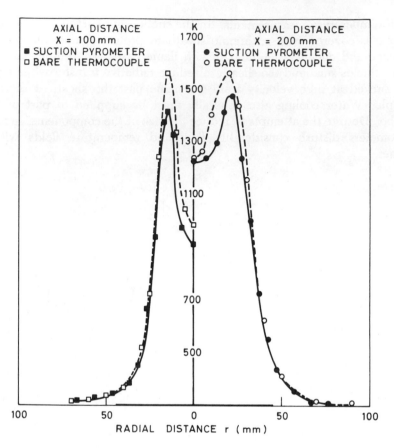

*Figure 4. Comparison of temperatures measured by suction py-
rometer and by coated thermocouple*

centrations, and the effects of radiative heat transfer from the probe to
the 'cold' laboratory surrounding were at a maximum.

The suction pyrometer became blocked with carbon particles, resulting in a decreased flow rate of gases around the thermocouple. This led
to poor convective heat transfer and caused significant changes in the
calibration of the instrument. As deposition increased with time, it was
difficult to compensate or allow for it. In measurements with the coated
thermocouple, carbon was seen to form on the bead. This increased the
insulation of the bead from the gases, thereby reducing heat transfer by
convection while, at the same time, increasing the emissivity of the
thermocouple and, hence, the radiation from the bead to the cooler surroundings. A detailed study of the components of conductive, convective,
and radiative transfer from the thermocouples to the surroundings as a
function of time and changes in the ambient conditions was not under-

taken but errors, based on simplified calculations, were of the order ±50°C in the high-temperature and highly turbulent regions of the spray flame. It could be assumed, however, that under the particular conditions of measurement by the two probes, the probe measuring the higher of the two temperatures was the more accurate. Since, in addition, the size and, hence, disturbance of the coated thermocouple was several orders of magnitude smaller than that of the suction pyrometer, it was concluded that the coated thermocouple was both the preferred and the more accurate of the two instruments.

The measured profiles in Figure 4 confirm previous measurements, showing a marked resemblance to temperature distributions in a gaseous diffusion flame. The main reaction zone, where temperatures reach peak

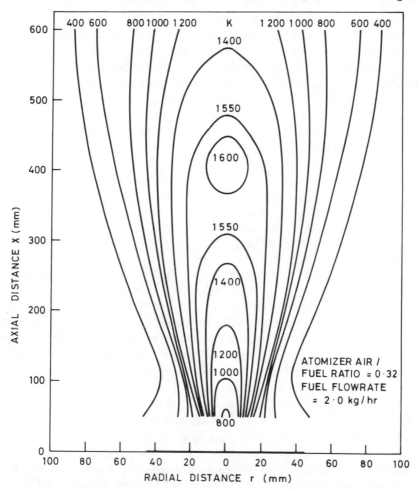

Figure 5.   Isotherms in liquid spray flame with disk stabilizer

values, is off the axis, and radially there are steep gradients of the order of 50 K/mm. Isotherms in the spray flame are shown in Figure 5. There is a cool central region lying within the high-temperature zone near the spray boundary. Along the centerline of the flame, temperatures rise as a result of entrainment of hot gases from the surrounding reaction zone until they reach a maximum at approximately $X = 400$ mm. Subsequently, temperature decreases on the axis because of entrainment of cooler air from the surroundings.

Concentration Measurements

Measurement of gas concentrations within flames provides information related to the rates of evaporation, mixing of reactants, reaction rates of combustion intermediates and products, and rates of dilution of combustion products. Quartz microprobes were constructed from right-angled 1-mm bore tubing, tapered at one end to produce a sonic throat orifice. Fristrom (6, 7) has shown that such probes provide minimum flow disturbance, low thermal conductivity, and sufficient quenching to give reproducible results within 3%. Fristom (7) has calculated that aerodynamic disturbances caused by the size of the probe and the withdrawal of the sample are of the order of 10% at 0.1 mm upstream for a 100-μm probe orifice. Errors can arise from the effects of thermal diffusion, catalysis, and quenching. The major sources of error for the batch sampling process when using gas-phase chromatography are:

(1) Inefficient quenching from incorrect fabrication of the quartz microprobes, resulting in continuation of chemical reaction along the sampling line. Properly designed probes may be 98% efficient.

(2) The absorption of combustion gases by condensed water vapor on transfer line walls.

These errors are minimized by reducing the length of the sampling line and maximizing the suction velocity. A peristaltic pump with a neoprene sampling tube provided a 10:1 pressure drop and was connected to a large vacuum pump to produce higher degrees of vacuum and increased flow rates. Pressure conditions in the sampling probe were between 10 and 30 mm Hg of vacuum. Radial traverses were made across the flame at various axial stations, providing a series of time average point concentration measurements.

Figure 6 shows radial profiles of concentrations of H_4, H_2, O_2, CO, and CO_2. Comparison of the profiles at axial distance $X = 80$ mm shows that CH_4 and H_2 have been formed from vaporization and cracking of the fuel vapor. At the spray boundary, $r = 9$ mm, O_2 concentrations are of the order of 3%, and, under these rich mixture conditions, reaction has taken place with CH_4 and H_2 to form CO. The peaks of CH_4, H_2, CO, and the minimum value for O_2 at $X = 80$ mm coincide with the re-

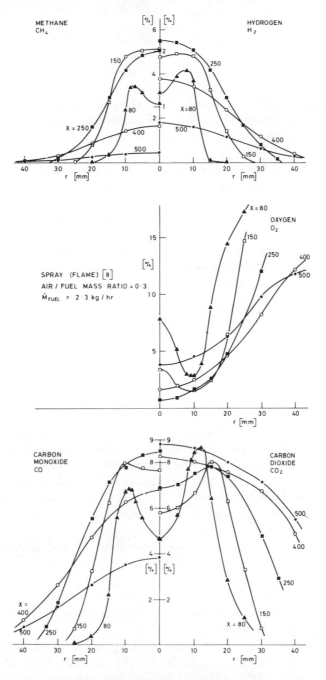

Figure 6. *Radial profiles of concentration of* CH_4, H_2,
O_2, CO, *and* CO_2 *at axial locations between* $X = 80$ *and*
$X = 500$ *mm*

gion of maximum temperature, as in Figure 5, which shows conditions at the flame front. Concentration variations along the centerline of the flame, $r = 0$, show that the flame can be subdivided into an initial region ($X < 250$ mm) followed by a flame region. In this initial region, mean temperatures (Figure 4) increase, but temperature levels are not sufficient to allow significant reaction under the rich mixture ratio conditions. Oxygen levels are caused by air coming directly from the atomizer orifice. Oxygen in the entrained air from the surroundings is mainly consumed in the peripheral flame region.

Methane and hydrogen concentrations increase with axial distance on the axis of the flame as a result of vaporization and cracking; thereafter these concentrations decrease as a consequence of reaction and dilution. The radially displaced peaks converge onto the flame centerline after $X = 250$ mm. At $X = 250$ mm, $r = 0$, and O_2 concentrations are less than 1%, indicating a region of intense combustion. The increases in O_2 concentration downstream of $X = 250$ mm are further evidence of increased entrainment of air from the surroundings. As O_2 concentrations increase, CO reacts with O_2 to complete the oxidation process by the formation of CO_2. Locations of peak values of temperature and O_2 coincide, indicating the downstream extremity of the combustion region, where dilution with air results in the nearest approach to overall mean composition corresponding to stoichiometric combustion.

Droplet Size Measurements

The measurement of the diameter of droplets in a spray provides information on the initial condition at the atomizer exit and on break-up or coagulation of droplets and vaporization. Fuel contained within droplets must go through the heating, vaporization, and mixing processes before it is available to react with oxygen, and these rate processes can be determined from the rates of decrease in the diameter of individual droplets.

Previous measurements of variation in droplet size in spray flames have been carried out by flash microphotography (3). In this study, variation in droplet size was measured using a recently developed laser diffraction meter (8). This technique provides the average size of droplets and the size distribution across the entire spray for a height equal to that of the laser light beam.

The laser diffraction meter consists of a parallel monochromatic light beam, 7 mm in diameter, from a 5-mW helium–neon laser, transmitted across the spray. Light diffracted by droplets and particles produces a Fraunhofer diffraction pattern. Light from the diffraction pattern is collected by 31 semicircular photosensitive rings, and the light energies across the pattern are derived from the measured output voltages. With the aid of a computer, the droplet size distribution is determined by fitting the experimental data to a Rosin–Rammler distribution:

$$R = \exp (-d/\overline{d})^N \tag{5}$$

where R = weight fraction of droplets larger than size d, $\overline{d}$ = characteristic mean size such that 63.2% of the droplets are less than $\overline{d}$, and N = measure of the spread of the size distribution.

Measurements in the isothermal spray using the laser diffraction meter are shown in Figure 7. Mean droplet diameter, as a function of

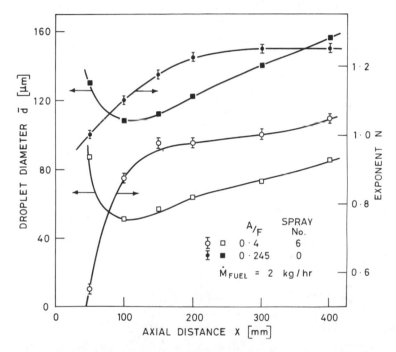

Figure 7.   Mean diameter of droplets and spread of size distribution N in isothermal spray. Measurements were made by laser diffraction meter across the spray at various axial locations.

axial distance, decreases between the atomizer exit and $X = 100$ mm as a result of the break up of large drops into smaller droplets. Further downstream where $X > 100$ mm, mean droplet diameters increase as a result of preferential vaporization of smaller droplets and relative effects of acceleration and deceleration on droplets according to their size.

The measure of spread, N, of the size distribution is also shown in Figure 7. At the exit of the atomizer, there is a high degree of poly-

dispersion with $N < 1.0$. The break-up of large drops into smaller droplets and the preferential vaporization of smaller droplets result in a
narrowing of the size distribution and, hence, in an increase in the value
of N. Figure 7 shows that when the fuel flow rate is constant and the air
flow rate is increased, both the mean droplet diameter and the values of
N decrease. The reduction in mean droplet size is caused by the increase
in shear forces between air and droplets as a result of the increase in
atomizer air exit velocity. The attempts to use the laser diffraction meter
in the spray flame were not successful, because of the influence of radiation from the flame. It is expected that, with the use of higher powered
lasers and filters to reduce the background radiation, measurements will
be possible in sprays under combustion conditions.

Conclusions

On the basis of measurements of velocity, temperature, and gas
concentration in spray flames and droplet size measurements in isothermal
sprays, the following conclusions have been reached.

Measured velocities in spray flames were higher than those in isothermal sprays. The acceleration of the flow is caused by increases in
volumetric flow rate as a result of an increase in temperature and the
change from liquid to vapor phase.

Reverse flow measurements in the wake of a stabilizer disk show
increases in size and magnitude of the maximum reverse flow velocities
within the recirculation zone as a result of combustion. Evidence was
found of initial pilot burning in the recirculation zone from combustion
of fine droplets transported by the reverse flow.

Temperature profiles in spray flames were similar to those previously
measured in gaseous diffusion flames. Measurement of temperature by
coated thermocouple was more accurate than measurements by suction
pyrometer within the temperature range and particular conditions of the
spray flames investigated. Changes in temperature within the flame
could be explained in terms of convection, reaction, entrainment, and
dilution.

Changes in concentration of gases are explained in terms of vaporization of liquid droplets, cracking of fuel vapor in fuel-rich gases, formation of intermediates, and, subsequently, completion of oxidation, followed by dilution of products by air.

Measurements of mean size and size distribution of droplets in
isothermal sprays by the laser diffraction meter show break-up of large
droplets into smaller droplets, followed by preferential vaporization of
smaller droplets, resulting in increases in mean droplet diameter and

reduced spread of the size distribution. Increases in atomizing air flow rates reduced mean droplet diameters.

Acknowledgments

The authors wish to thank the Atomic Energy Research Establishment, Harwell, for financial support during the investigation and G. Wigley for the loan of the miniature suction pyrometer.

Literature Cited

1. Chigier, N. A., "The Atomization and Burning of Liquid Fuel Sprays," *Prog. Energy Combust. Sci.* (1976) **2.**
2. Hetsroni, G., Sokolov, M., "Distribution of Mass, Velocity and Intensity of Turbulence in a Two-Phase Turbulent Jet," *Trans. ASME, Ser. E. (J. Appl. Mech.)* (1971) **38**, 315–327.
3. McCreath, C. G., Roett, M. F., Chigier, N. A., "A Technique for Measurement of Velocities and Size of Particles in Flames," *J. Phys. E* (1972) **5**, 601–604.
4. Chigier, N. A., Dvorak, K., *Symp. (Int.) Combust. 15th (Proc.)* (1975) 573.
5. Onuma, Y., Ogasawara, M., *Symp. (Int.) Combust. (Proc.) 15th* (1975) 453.
6. Fristrom, R. M., Grunfelder, C., Favin, S., *J. Phys. Chem.* (1960) **64**, 1386.
7. Fristrom, R. M., Prescott, R., Grunfelder, C., *Combust. Flame* (1957) **1**, 102.
8. Swithenbank, J., Beér, J. M., Taylor, D. S., Abbott, D., McCreath, C. G., 14th Aerospace Sciences Meeting, AIAA Paper No. **76-69**, Jan. 1976.

RECEIVED December 29, 1976.

8

Applicability of Laser Interferometry Technique for Drop Size Determination

ERIC W. SCHMIDT, ANTHONY A. BOIARSKI,
JAMES A. GIESEKE, and RUSSELL H. BARNES

Battelle, Columbus Laboratories, 505 King Ave., Columbus, OH 43201

An experimental research program to determine the feasibility of measuring drop sizes in a spray environment using a laser interferometry technique showed that noninvasive real-time measurements of both droplet size and velocity can be made in sprays generated by a fuel atomization nozzle using an interferometry/visibility technique with a programmable calculating oscilloscope. Interferometric size determinations compared favorably with microscopic examination of drops collected on MgO-coated slides and also with Mie scattering measurements. The technique was sensitive enough to detect small changes in droplet size caused by evaporation and by changes in atomization characteristics.

Droplet size is an important controlling parameter in the combustion of sprays. Therefore, the ability to measure accurately droplet sizes in a spray environment is necessary if detailed studies are to be made of spray combustion. An experimental program was conducted to demonstrate the feasibility of using an interferometry technique to measure drop sizes in sprays generated by a fuel atomization nozzle, and this chapter discusses the accuracy and sensitivity of the technique.

Droplet size may be determined with the use of a laser interferometer. This technique is an extension of the laser Doppler anemometry (LDA) technique commonly used to measure velocities of small particles in a flowing stream. Two equal-intensity Gaussian beams are made to intersect at their focal waists to form a standing electromagnetic wave distribution in the intersection (probe) volume. This standing wave can be visualized conceptually as a set of planar fringes perpendicular to the plane of the two beams and parallel to their bisector. When a light-scattering particle, such as a small liquid droplet, passes through the probe volume, the scattered intensity varies with time. The inverse of this time periodicity is termed the Doppler frequency.

It has been suggested (1) that the Doppler frequency is the rate at which a droplet crosses alternate bright and dark interference fringes. Thus, exact knowledge of fringe spacing and measurement of the Doppler frequency allow a direct calculation of droplet velocity. Likewise, it has been shown (2) that analysis of the AC and DC components of the scattered light resulting from a droplet crossing alternate bright and dark fringes gives a droplet visibility which can be correlated to droplet size.

Theory

The visibility function, V, is defined as the ratio of the AC to DC components of the observed scattered signal from the interference fringes. It is fully equivalent to the Michelson visibility function and may be defined as:

$$V = \frac{I_{max} - I_{min}}{I_{max} + I_{min}} \tag{1}$$

where I_{max} = maximum scattered intensity from a bright fringe, and I_{min} = corresponding minimum scattered intensity from the next consecutive dark fringe.

The visibility function can be reduced to a droplet size parameter (droplet diameter/fringe spacing) analytically. The appropriate equations and their analysis have been presented by Farmer (2). He has shown that under certain constraints, V reduces to:

$$V \simeq \frac{2J_1(\pi D/\delta)}{(\pi D/\delta)} \tag{2}$$

where D is the droplet diameter, δ is the fringe spacing, and J_1 is a first-order Bessel function with argument $\pi D/\delta$. The constraints under which Equation 2 applies are:

(1) Mie scattering theory is valid

(2) Crossing angle between the two beams is small

(3) Equal intensity Gaussian beams

(4) Polarization vectors of two beams are parallel to each other and are perpendicular to the plane of intersection

(5) Observation angle of collector is near 0° or 180° (i.e., small angle forward or back scattering)

(6) Small viewing aperature is required such that Mie scattering cross section does not vary over the collector solid angle

(7) Spherical scattering particles

(8) Single observed scattering event (i.e., only one scattering particle in probe volume at time of measurement

(9) Particle passes through the center of the beam crossing volume to prevent fringe contrast variations

(10) Particle diameter is much less than diffraction-limited beam focal diameter

These restrictions are required in order to reduce the complex formulation obtained from the general theory of Mie scattering to the simple expression given in Equation 2. This expression is plotted in Figure 1. As shown in the figure, the multinodal nature of visibility function can lead to ambiguous results for low visibilities. In order to avoid this problem, the visibility technique should be used only when the droplet-size-to-fringe spacing ratio lies within the first node of the function. More precisely, to be completely unambiguous, the diameter-to-fringe spacing ratio, (D/δ), must be less than 1.05, and hence the visibility will be greater than 0.13 as indicated in Figure 1. For conditions that violate the above constraints, Orloff (3) found that the region of ambiguity extended to even higher visibilities of the order of 0.5. In any case, adherence to the many constraints outlined above should be checked for each experimental setup by measuring the visibility function for a range of known particle sizes.

After the visibility indicated by a certain signal has been determined, droplet size can be obtained more expeditiously if a simpler equation

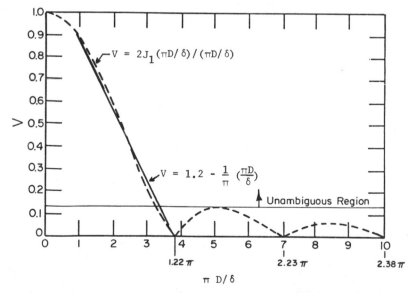

Figure 1.   Fringe visibility vs. particle size for a sphere

than the Bessel function relation of Equation 2 is used to relate size to visibility. A linear equation was adequate for this purpose. This relation is:

$$V = 1.2 - D/\delta \tag{3}$$

Between the limits of $0.3 \leq D/\delta \leq 1.05$, this linear equation models the Bessel function relation to within 5% throughout the region. Figure 1 shows the comparison between the linear relation and the previously noted visibility function. Because of the linear approximation, there is an obvious limitation in that D/δ must be within a certain region; however, for $D/\delta < 0.3$, the visibility function is highly insensitive to diameter variation, and, thus, this particular region (i.e., $V > 0.9$) is not well suited for practical application. Further, the linear equation limit of $D/\delta \leq 1.05$ is not prohibitive since this D/δ value nearly corresponds to the visibility function ambiguity limit of $V > 0.15$, and this region should be avoided. Hence, the restrictions placed in the linear relationship given in Equation 3 are compatible with the restrictions placed upon the visibility technique itself. Also, the simplicity of Equation 3 makes it very useful for reducing large amounts of data efficiently.

Experimental Apparatus and Procedures

Spray Generation. The fuel spray analyzed in this study was generated from a two-fluid nozzle. No. 2 diesel fuel or kerosene was the primary fluid, and compressed air was the atomizing fluid. The fuel-to-air mass flow ratio was varied to obtain different atomization characteristics and, therefore, to alter droplet size. The atomizing nozzle was located in a cylindrical duct through which room air was drawn. The nozzle assembly was mounted on a traversing system and was moved axially with respect to the probe volume in order to make axial transverses without disturbing the laser droplet measuring system. Measurements of the spray drop sizes were made at the spray and duct centerline along an axial traverse in order to examine drop size variations resulting from evaporation. Figure 2 shows the flow system used for spray generation.

Diagnostics and Data Processing. A schematic of the laser diagnostic equipment used to perform the interferometer/visibility measurements is shown in Figure 3. A 1-W argon-ion laser was used to provide a 200-W monochromatic beam at 4880 Å. This beam was turned 90° and was directed into the beam splitter by an adjustable mirror. A variable attenuator in one beam path within the module was adjusted in order to obtain equal beam intensities of approximately 50 mW. The emerging beams, initially 20 mm apart, were focused and made to intersect at an angle of 1.91° using a 600-mm focal length lens. This angle

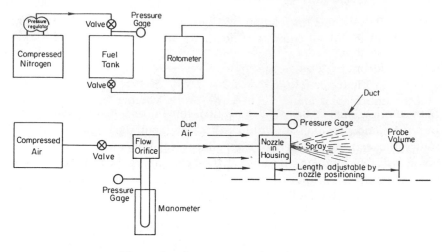

Figure 2.   Spray system flow diagram

would result in a 14.7-μ fringe spacing. A second lens was added to obtain smaller beam crossing angles, and hence, larger fringe sizes, by forming another crossing point inside the liquid spray jet. The focal length of this second lens was 120 mm. By locating the first beam crossing position a prescribed distance from the second lens, a much smaller crossing angle of 0.314° was obtained. This dual lens system provided a maximum fringe size of 89 μm. Also, by adjusting the initial beam separation to 40 and 80 mm, one could obtain 44- and 22-μm fringe spacings. This setup was flexible enough to allow measurements of droplet diam-

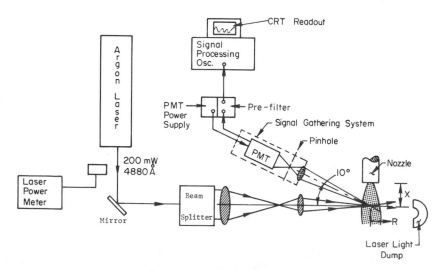

Figure 3.   Block diagram of apparatus used in atomization tests

eters from 5 to 94 μm. However, one fringe spacing, 89 μm, was adequate for the drops produced in this study.

Light scattered from droplets within the fringe region was collected and focused onto a pinhole in front of a photomultiplier tube (PMT) as shown in Figure 3. The intersection of the laser beam path with the collector optics solid angle defined the probe volume or measurement volume. Since the collector was located at a 10° angle from direct back-scatter, the probe volume was quite small. Calculations indicate a somewhat cylindrical probe volume 0.2 mm in diameter and 6.0 mm long. This volume was much smaller than would have been obtained using direct backscatter. The small probe volume dimensions proved to be useful in high-density regions near the atomizer nozzle exit.

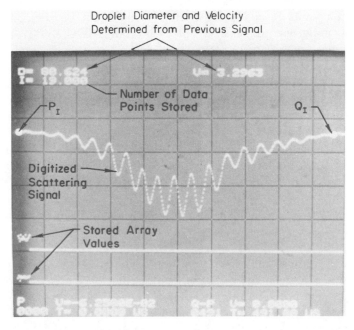

Figure 4.   Cathode ray tube display of digitized droplet scattering signal

The PMT signal was prefiltered to remove the DC dark current level of the tube and then digitized by a signal-processing oscilloscope. A typical digitized signal display is shown in Figure 4. Each signal was either accepted or rejected by the investigator based on its uniformity and periodicity which indicated whether or not the signal represented a single particle scattering event. The accepted signals were then processed using a Norland model 2001 programmable calculating oscilloscope to obtain the droplet size (diameter) and velocity based on the visibility and

Doppler frequency of the signal, respectively. For any given experimental condition, a number of measurements were made and stored. These stored parameters were then analyzed using the same programmable calculating oscilloscope to give both a statistical analysis (arithmetic mean and standard deviation) and a representation of the data in the form of a histogram.

Some advantages and disadvantages of the above processing technique should be noted. The trigger function for this continuous memory digital system provides for recording and display of signal which occurred prior to the trigger event. Hence, the trigger level need not be near zero to produce entire pedestal waveform as shown in Figure 4. Another advantage is the floating zero feature of the data analysis. By floating zero it is meant that the actual voltage of the no-scattering position on the waveform (points P_I and Q_I in Figure 4) need not be zero. This was especially important in dense droplet flows since DC shifts can occur from the close proximity of other scattering events. The signal analysis did not require a zero reference. So, the relative magnitude of each point in the array was referenced to a no-scattering point ahead of the pedestal waveform rather than to zero. The greatest disadvantage of this data analysis technique is the calculation speed which could be increased only by going to a mini-computer system.

Experimental Comparison with Other Techniques. In order to check that the experimental setup shown in Figure 3 met the constraints formulation given in Equation 2, two alternate techniques were used to measure droplet size during selected tests. These were a magnesium-oxide (MgO) coated slide collection method and a single-beam Mie scattering technique.

MAGNESIUM OXIDE METHOD FOR DROP SIZE MEASUREMENT. Droplet size may be determined from the impression made by a drop when collected on a coated slide. A layer of MgO has been used commonly as a collecting medium. Drops form a crater in the MgO on impaction. Based on past calibrations (3, 4), the diameter of this crater, as determined from microscopic observation, can then be related to droplet size.

In making collections in a fuel spray for this study, two potential problems were avoided. First, because of the density of the spray, short exposure times were required. Secondly, in order to make accurate comparisons, data were desired only along the centerline with no interfering edge effects. Both of those problems were overcome by covering the coated slide during insertion into the spray, exposing it momentarily to make the impaction collection, and then covering it once again before removal.

MIE SCATTERING TECHNIQUE FOR DROPLET SIZING. Single-beam laser light scattering measurements were performed to determine average droplet size. For these measurements, a photomultiplier system was attached to an arm which rotated about a pivot located beneath the spray so that scattered light intensity from the argon-ion laser beam could be recorded continuously as a function of angle from the forward direction of the beam. Measurements were made in both the back-scattering (9°–70°) and forward-scattering (115°–180°) direction. The physical setup made measurements between 70° and 115° impossible, however, this gap contains no important information for the droplet size range involved in this study.

Relative scatter intensity as a function of scatter angle was measured, and the experimental curve was fit to curves calculated from Mie theory (5, 6) in order to determine average droplet size. In the present work where large-diameter droplets are involved (diameters greater than a few times the wavelength of the light), the Mie scattering equations reduce to a simple diffraction relationship. In this case, the scattering intensity, $I_o(\theta,d)$, which is independent of refractive index, can be expressed as (7):

$$I_o(\theta,d) = \frac{J_1^2(\alpha \sin \theta)}{\pi \sin^2 \theta} \left[\tfrac{1}{2}(1 + \cos^2 \theta) \right], \qquad (4)$$

where $\alpha = \pi d/\lambda$, d = particle or droplet diameter, λ = wavelength of light, θ = scattering angle measured from the forward direction, and J_1 = Bessel function of first order.

To determine the average droplet size from the scattering data, theoretical curves of the ratio of $I(\theta,d)/I(4°,d)$ were calculated as a function of θ for a range of appropriate droplet diameters.

Results

Comparison of the Visibility/Interferometer Technique with the MgO Collection Technique. During one experiment, using kerosene as the primary fluid, a MgO-coated microscope slide was used to collect droplets 40 in. from the nozzle. The impressions formed by the droplets in the MgO were then viewed and sized using an optical microscope. During the same experiment and at the same location, drop sizes were measured using the visibility technique. One hundred drops were analyzed using each technique, and comparisons were drawn.

Figure 5 shows the respective distributions of the two techniques plotted as cumulative percent by number against size. The two curves appear quite similar. The number median of the MgO data is 60 μm. This result represents a less than 4% difference, which is a very favorable comparison. The MgO analysis indicates a broader distribution. This may be attributed to the limited range of the visibility technique as previously noted. Owing to this limitation, any droplet larger than 94

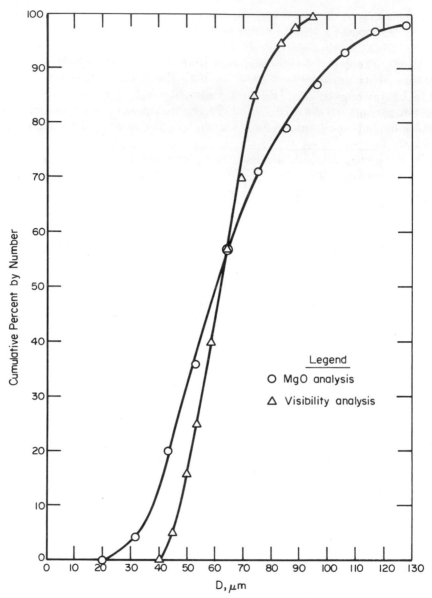

Figure 5.   Size distribution of kerosene spray 40 in. from nozzle

μm diameter (for this experiment $\delta = 89$ μm, thus the upper limit of 1.05 δ is approximately 94 μm) will be interpreted as a 94-μm particle. Similarly, detection of small drops with the visibility technique is limited theoretically to drops larger than about 30 μm ($D/\delta = 0.3$). It appears, however, that a more practical limit for this fringe spacing is about 40 μm. This bias at the lower end may be caused by exceeding the limit of the linear modeled visibility function used or by the signal triggering effects from lesser amounts of scattering seen from relatively small droplets. These problems could be overcome by performing visibility measurements using several additional fringe spacings. Over the range of technique applicability it must be concluded that differences between the two techniques are quite small. Possibly a clearer comparison of the data obtained by the two techniques is shown by Figure 6. Again a

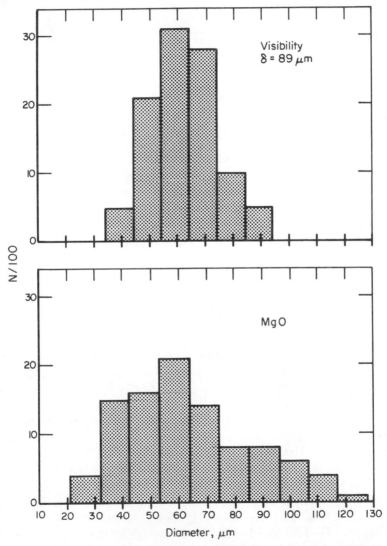

Figure 6.   Comparison of droplet size distributions—MgO and visibility techniques

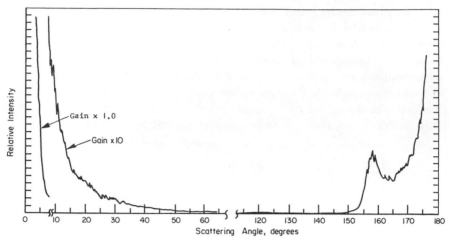

Figure 7.   Laser light scattering at 4880 from A from kerosene spray at 10 in. from the atomizing nozzle

comparison reveals a similar mean diameter (the modes of the two histograms appear in the same range) with the broader distribution of the MgO data as noted. This comparison of size distribution data by two independent techniques is quite favorable and indicates the accuracy and applicability of the visibility technique.

Comparison of the Visibility/Interferometer Technique with the Mie Scattering Technique. Droplet size measurements were also made based on the total light scattered from a single beam at various angles. During one experiment, measurements were made at several axial distances from the nozzle using the Mie scattering technique, however, only one point (at 10 in. from the nozzle) was analyzed because of the time-consuming effort involved in data reduction. The experimental curve of this measurement is shown in Figure 7. A Mie scattering analysis of this curve indicates a mean drop diameter of 70–75 μm. This compares quite favorably to a 79-μm mean diameter measured using the visibility technique.

In comparing the two methods, the angle technique requires a less elaborate laboratory setup; however, the additional amount of time and effort involved in data reduction, especially for droplets larger than 10 μm, makes this technique much less desirable than the real-time visibility technique. Furthermore, unlike the visibility technique, the Mie scattering technique gives only a mean size rather than a complete size distribution.

Capability to Measure Small Changes in Droplet Size. Several experiments were performed to assess the capability of the interferometer/visibility technique to measure drop size changes resulting from changes in the atomization procedure or from subsequent drop evaporation. Using kerosene as the primary fluid, droplet evaporation rates were altered by making slight changes in fuel and environment temperature. Experiments A and B were conducted with all conditions remaining constant except that during Experiment B, the fuel and duct air temperatures were about 30°F higher. Even with this slight change, differences in droplet size at various axial points resulting from evaporation were noticed in the results from the two experiments. Figure 8 shows greater initial evaporation at the higher temperature conditions.

Two experiments were conducted using No. 2 diesel fuel atomized at different air/fuel ratios to show that changes in droplet size caused by atomization can be measured. During the first of these experiments the air/fuel ratio was set at 1/1 while the second was conducted with an air/fuel ratio of 2/3. All other conditions were identical. The measured mean droplet diameters 40 in. downstream of the nozzle were 76.5 and 84.8 μm, respectively, thus indicating better atomization during the first experiment resulting from the higher air/fuel ratio.

Conclusions

The 10° backscatter angle used in the present experimental setup tends to violate the coaxial scattering assumption used to obtain Equation 2. However, the final results of our experiments showed that this non-coaxial detection provided accurate measurements of kerosene spray drop size and velocity in the 50–90-μm diameter drop size range. Further,

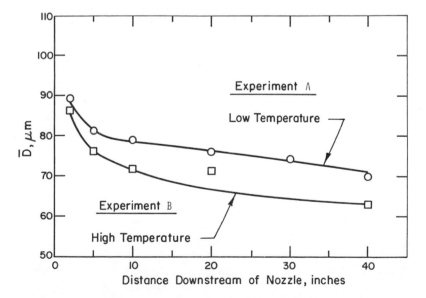

Figure 8.   Kerosene droplet size results

the results demonstrated that the interferometric/visibility technique can detect small changes in droplet size and velocity, thus making it possible to monitor atomization and evaporation as dependent on atomization schemes and conditions and on air flow conditions. The technique also has the advantage of being a real-time and noninvasive measurement. Signal analysis methods are available to accumulate and analyze data in such a manner that the desired screening, averaging, and statistical analyses of the data may be performed in almost real time.

Literature Cited

1. Rudd, M. J., *J. Phys.* (1969) **E 2**, 55.
2. Farmer, W. M., *Appl. Opt.* (November, 1972) **11**.
3. Orloff, K. L., private communication.
4. May, K. R., *J. Sci. Instrum.* (1950) **27**.
5. Gieseke, J. A., Mitchell, R. I., *J. Chem. Eng. Data* (October, 1965) **10** (4).
6. Van der Hulst, H. C., "Light Scattering by Small Particles," Wiley, 1957.
7. Kerker, M., "The Scattering of Light," Academic (1969).
8. Hodkinson, J. R., Greenleaves, I., "Computation of Light-Scattering and Extinction by Spheres According to Diffraction and Geometrical Optics, and Some Comparisons with Mie Theory," *J. Opt. Soc. Am.* (1963) **53**, 577.

RECEIVED December 29, 1976.

Combustion By-products

Ionization Associated with Solid Particles in Flames

R. N. NEWMAN, F. M. PAGE, and D. E. WOOLLEY

University of Aston in Birmingham, Gosta Green, Birmingham, England

Ionization associated with the presence of solid particles in flames is small and easily obscured by gaseous ionization of volatile impurities. It can be demonstrated experimentally in certain systems and can be shown to depend on the particle size, number density, and work function as predicted by the theory of Smith or Soo and Dimick. Salts such as the alkali halides volatilize slowly and mix by diffusion. Residual inhomogeneities in ion distribution give the appearance of particulate ionization.

A flame generally conducts electricity by the free electrons present at high temperatures. These electrons may be derived from the thermal ionization of substances present in the flame, e.g., an alkali metal (1) or may be produced chemically in ionization reactions such as those between CH radicals and oxygen atoms (2). The factors which control the concentration of free electrons in these homogenous processes may be analyzed by classical physical chemical arguments and are well understood. Unfortunately, real flames depart from the ideal systems to which these arguments apply and are frequently heterogenous, containing solid particles or smokes. If this occurs, free electrons may be produced by thermionic processes, and the analysis of the electron concentration then depends on the detailed nature of the smoke particles. The experimental investigation of suitable systems over the past few years has shown that such analysis is possible but that under most normal conditions the homogenous processes dictate the levels of ionization observed.

The flames which have been used in these studies are premixed laminar hydrogen/nitrogen/oxygen flames supported on a Meker burner and screened by a second flame of similar composition. These flames are well characterized, and the temperature and composition are well known. In addition, the transient concentrations of free radicals and other minor

0-8412-0383-0/78/33-166-141$05.00/0 © 1978 American Chemical Society

species have been measured experimentally by spectroscopic techniques. The flames are thus effectively inert isothermal high-temperature matrices in which to study the ionization processes but have spatial radical profiles which may be used to trace chemical effects. The design of burners and the techniques of flame characterization are described fully elsewhere (3, 4, 5).

The clean flames show low levels of ionization close to the reaction zone because of chemi-ionization processes such as:

$$CH + O \rightarrow CHO^+ + e$$

arising from traces of hydrocarbons in the gas supply. This ionization decreases rapidly with height through recombination and is negligible 5 cm downstream where the measurements were made. If the vapor of an ionizable material is added to the clean flame, a measurable level of ionization results, arising either from direct thermal ionization if the ionization potential is low:

$$Cs + M \rightarrow Cs^+ + e + M$$

or by charge transfer from the hydrocarbon ion if the ionization potential is high:

$$Pb + CHO^+ \rightarrow Pb^+ + CO + H$$

The resultant atomic ions recombine with electrons slowly, so that a high, kinetically controlled level of ionization may result.

The ionization processes have been studied extensively by microwave absorption to detect electrons by electrostatic probes which may be biased to detect either negative species (almost entirely electrons) or positive ions, and by mass spectrometry, which has shown that molecule ions such as $SrOH^+$ are common (6).

As a result of these studies, a considerable and coherent body of information exists about the ionization kinetics, energetics, and processes for many volatile additives. The presence of solid particles in the flame has always been associated with ionization, and this chapter describes some attempts to make and interpret such observations.

Addition of Particles to Flames

An atom is about 0.33 nm in diameter, so that a particle whose diameter is 0.1 μ will contain about 3000^3 or 10^{10} atoms. Any observations of the ionization associated with such a particle must take this into account, both because of the large amount to be added to achieve a given particle number density and because of the relative importance of traces of vapor of the material itself or of impurities. There are three principal ways to add particles to a flame:

(1) Directly as a dust, a method which is not easy to control and which is limited to relatively coarse particles,

(2) As the spray of a solution, which is easy to control but where the size of particle rather than the number density is varied, and

(3) Autogenously, whereby the vapor of a precursor to the particle is mixed uniformly with the flame gases.

This last method includes soot formation in fuel-rich hydrocarbon flames. All methods have been used successfully, and the presence of ionization from particles is demonstrated convincingly.

The appearance of the flame is no guarantee that particles are present. Iron carbonyl produces a luminous flame typical of particulate matter, but particles, if present, are no more than a few atoms across, and the luminosity results from overlapping molecular band spectra. The ionization produced follows a square root dilution law typical of free atoms.

Theory of Ionization of Solid Particles

The first attempts to describe the ionization of a flame containing particles was made by Sugden and Thrush (7) who modified Richardson's equation (Equation 1) for the vapor pressure of electrons over an infinite plane surface:

$$n_e = \frac{2(2\pi m k T)^{3/2}}{h^3} \exp(-\phi/kT) \tag{1}$$

where n_e is the number density of electrons and ϕ the work function of the surface, to take account of the accumulation of positive charge on a finite, spherical particle by adding a term Ze^2/a where Z is the number of charges on the particle and a the radius. Since, in the absence of negative ions, n_e may be put equal to Zn_p, where n_p is the number density of particles and Z is now their mean charge, the exponential in Equation 1 may be written as:

$$\exp[-\{\phi + (n_e/n_p)e^2/a\}/kT] \tag{2}$$

This argument has been developed further to include the effects of charge distribution and of other gaseous ionization, and the final equations of Smith (8) agree generally with an alternative kinetic approach such as that of Soo and Dimick (9). It is convenient to express the electron number density in terms of the saturated number density over an infinite plane lattice (n_s) so that:

$$n_e/n_s = exp\,(-\,\{n_e + [B^-] - [A^+] + \tfrac{1}{2}\}\,e^2/an_pkT\,) \qquad (3)$$

where A^+ and B^- are the gaseous ions, and the additional factor of $\tfrac{1}{2}$ arises from the interaction of positive ions and electrons. This relation can be put in the form:

$$n_e = \frac{an_pkT}{e^2}\ln\left(\frac{n_e}{n_s}\right) - \frac{n_p}{2} + [A^+] - [B^-] \qquad (4)$$

or in a form retaining the work function of the surface as:

$$\ln\,(n_e/AT^{3/2}) + (e^2/2akT)\,(1 + 2n_e/n_p) = \phi/kT +$$
$$\{[A^+] - [B^-]\}\,(e^2/n_pakT)$$

If the observed ionization is caused in large part by the solid particles, these equations may be simplified by omitting the terms in $[A^+]$ and $[B^-]$, but if gaseous ions are present, either added deliberately or produced by volatilization of the solid, this may not be done. Quite massive quantities of strong electron acceptors are required before the negative ion concentration becomes significant, so that $[B^-]$ rarely becomes important.

Experimental Results

Particle Size. Uranium salts, when sprayed into a flame as an aqueous solution, give a smoke which evaporates at temperatures above 2550 K but which persists evenly throughout the flame at 2500 K. Ionization could be measured readily at 2500 K where little gaseous uranium could be present. Each solid particle of the smoke arises from the evaporation of a single droplet of solution, and the weight of each particle will be determined by the weight dissolved in each droplet. The size distribution of the droplets is determined principally by the aerodynamics of the delivery system and does not depend on the concentration of salt in the solution. Altering that concentration will therefore alter directly the size of the ionizing particle, and in fact the radius of the particle should be proportional to the cube root of the molarity of the solution sprayed. In a constant flame (T constant) with no gaseous ionization ($[A^+] = [B^-] = 0$) and with a constant particle number density, Equation 4 indicates that the electron number density (n_e) is proportional to the radius (a) provided that ionization is small so that $\ln\,(n_e/n_s)$ varies only slowly. These conditions are satisfied, and the predicted relation is

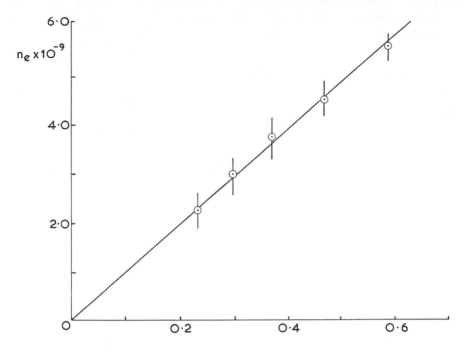

Figure 1. Dependence of the electron concentration on the cube root of the molarity of uranyl nitrate solution sprayed into a flame

shown in Figure 1. The particle number density cannot be changed by changing the concentration of solution sprayed, although the total amount of material in the flame does change.

The results from a series of flames at different temperatures are analyzed by Equation 5 in Figure 2 where the function $kT \ln (n_e/AT^{3/2})$ is plotted against n_e. The intercept (3.6 eV) agrees well with the literature value for uranium dioxide, and the slope gives a value of 3.3×10^6 m^2 for the product $n_p a$. The sprayer had been calibrated with cesium, and it was known thus that the total number density of uranium atoms in the flame was 1.6×10^{24} m^{-3} which corresponds to a relative volume $(4\pi n_p a^3/3)$ of 6.5×10^{-10} of UO_2 (density 10.97 g cm^{-3}). This leads to separate values for a (the mean particle radius $= 6.8 \times 10^{-9}$ m) and n_p (particle number density $= 4.8 \times 10^{14}$ m^{-3}) and to the number of charges per particle ($n_e/n_p = 7.2$).

If each smoke particle is derived from a crystallite of the sprayed salt, uranyl nitrate, the particles will be porous and the density much less. Such an assumption increases the mean radius to 18.5×10^{-9} m and the number of charges per particle to 20 while lowering the particle number density to 1.8×10^{14} m^{-3}. In either case the particles contain approximately 10^5 atoms.

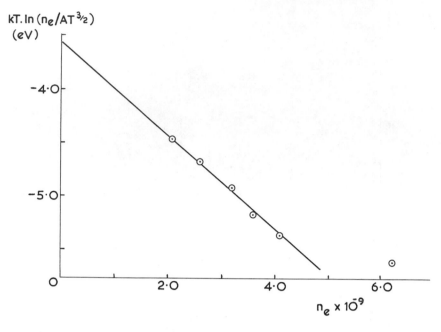

Figure 2.   *Determination of the work function of the UO$_2$ particles*

Effects of Gaseous Ionization

Whereas uranium dioxide is a relatively involatile material, nickel is not, but it will still produce a smoke associated with a detectable degree of ionization. Nickel may be added conveniently to the flame as the vapor of nickel carbonyl with the general results illustrated by Figure 3. The ionization at low partial pressure is ascribed to gaseous nickel, but the levels of ionization were too low to demonstrate with certainty that the ionization was proportional to the square root of the nickel concentration. Iron pentacarbonyl under similar conditions did give an accurate square root law behavior characteristic of atomic ionization (*10*). As the total amount of nickel in the flame was increased, the ionization rose sharply and reached a fresh line of proportionality where ionization was caused by the condensed phase. The accessible range of conditions was limited at low temperature by the effects of the added carbonyl and at high temperatures by evaporation of the smoke. However the work function was evaluated as 3.1 ± 0.5 eV (literature value, 4.0–5.2 for metallic nickel) with a mean particle radius of 21 nm and number density of 1.5×10^{15} m^{-3} and was proportional to the partial pressure of nickel carbonyl in the unburned flame gases. These particles could buffer electrons from added alkali salts—a CsCl solution sprayed into a flame gave 2.4×10^{16} m^{-3} of electrons alone, but when the flame contained

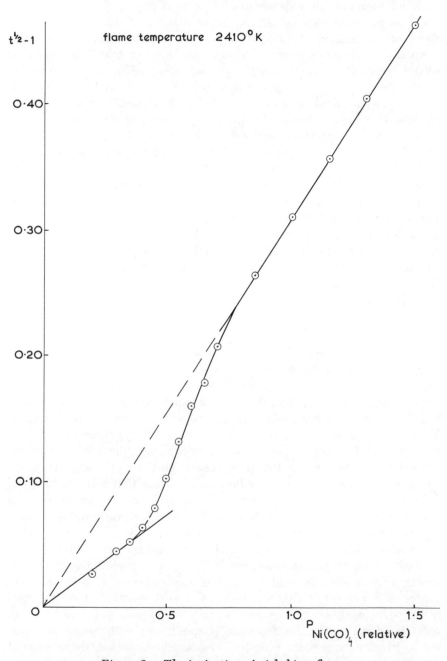

Figure 3. The ionization of nickel in a flame

nickel particles with an associated electron density of 5.6×10^{16}, the electrons from particles and cesium together amounted only to 6.7×10^{16}, a clear and quantitative demonstration of buffering.

The measurements on which Figure 3 is based were obtained in a flame of temperature 2410 K, at a position corresponding to 3.6 msec from the combustion zone. It would be expected that solid particles would begin to appear when the partial vapor pressure of the nickel reactant was 2.5×10^{-2} atm, but the onset of particle ionization is much earlier, at a partial pressure of 7×10^{-4} atm. Evidence is presented below to show that the average partial pressure, which is a total atom number density, has little meaning in the presence of solid particles. It is probable that the particles are formed in the preheating zone of the flame (11) and that their dispersal is controlled by diffusion. A particle number density of 10^{15} m^{-3} implies that each particle is 10 μm, or 200 particle diameters, from its neighbor. Each particle 20 nm in diameter will contain approximately 10^6 atoms, and if the onset corresponds to the particle just evaporating in 3.6 msec at the vapor pressure of 2.5×10^{-2} atm, a diffusion coefficient of 10^{-4} m^2 sec^{-1} would limit the vapor to 110 nm from the particle surface. The effect of partial pressure on the ionization behavior is·ascribed therefore to the size of particle formed in the preheating zone rather than to a limit for condensation.

Studies with Electrostatic Probes

Measurable levels of ionization from autogenously formed smokes were associated invariably with small particles, and an increase in particle size was sought through the direct addition of dusts. There were many difficulties in this, and the resonant cavity technique was not well suited to these measurements. Such dust-laden flames were examined therefore by electrostatic probes, which can handle particles up to 3 mm in diameter. The response of a probe moved through an ionized flame is normally smooth, but in the presence of particles it was very noisy (Figure 4). After many trials it became apparent that this was the real behavior of a probe in such a flame and that each spike during the traverse corresponded to the interaction between the probe and a particle. The general behavior corresponds to a cold probe which is non-emitting interacting with a hot particle which can emit but which is capacity limited. During the interaction, the hot particle is cooled, but the probe can emit thermionically until cooling has occurred or until the interaction ceases. The current flowing will reflect the work function of the material of the particle, although the temperature is ill-defined, and a quantitative correspondence not to be expected. Experiments with aluminum, alumina, barium oxide, and tungsten hexaboride bore out these general conclusions, but it is not possible to exclude the effects of volatile impurities causing associated gaseous ionization. Certain experiments directed to this possibility gave some interesting results.

The smooth curve which resulted when an alkali halide solution was sprayed into the flame occasionally showed slight noise. Since such a spray corresponds to the injection of crystal dust because the droplets evaporate before reaching the burner, the probe response should be the same as for any dust. The number of droplets entering the flame was reduced while keeping the droplet size constant, by replacing the spray of the atomizer by a bubbler and by leaving the settling column unchanged. Spray was then produced only by the bursting of bubbles, but the resulting drops which emerged from the atomizer, although few in number, each had the same size as those produced by the sprayer. The

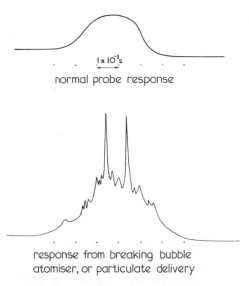

normal probe response

response from breaking bubble
atomiser, or particulate delivery

Figure 4. Oscilloscope traces from electrostatic probes passing through flames

use of this highly inefficient atomizer reduced the number of particles entering the flame to a level comparable with that achieved by direct addition and resulted in the same spiky traces.

Samples of the dust from a potassium chloride solution collected with no flame burning showed it to be composed of well formed crystals with no trace of surviving droplets and to be visually of a very narrow size distribution with a mean diameter of 4×10^{-6} m. The oscillogram spikes had a width-at-half-height of 200 μsec. There was no correlation between width and height, which suggests that the width is associated with gaseous ionization arising from the particle and not with the particle itself. This half-width corresponds to a distance of 5×10^{-3} m (flame gas velocity = 24 msec^{-1}).

The linear velocity of the probe tip (1 msec^{-1}) is low compared with the gas velocity, so that this distance must be measured in the direction of gas flow, and the full width of the spike can correspond only to the passage of 10×10^{-3} m of ionized gas over the probe. A crystallite of KCl with the observed diameter, if volatilized at 1 atm and 2500 K (the measured flame temperature) would produce a vapor sphere of 10^{-5} m radius or a cylinder 10×10^{-3} m long and 4.5×10^{-6} m in diameter. This suggests that the spikes are indeed caused by a trail of vapor left as the particle evaporates.

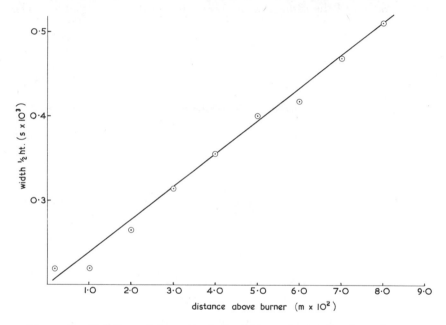

Figure 5.   Relation of the spike half-width to the height above burner

Normally, one would expect that a small particle would attain rapidly, but not instantaneously, the velocity of the surrounding gas and that during the acceleration the particle would move relative to the gases, leaving the type of trail observed. There are actually two regimes to consider. In the burner, the particles are carried through 37 fine tubes, 6×10^{-2} m long and 8×10^{-4} m i.d. The unburned gases in these tubes have a mean velocity of 17 msec^{-1}. The velocity gradient in the tube will carry the small particles to the center of the tube, and they will cross the flame front at a velocity of 34 msec^{-1} into the burnt gases moving at 2500 K at 24 msec^{-1}.

The drag coefficients for small particles have been tabulated (*12*) as a function of particle radius. By integrating the equation of motion, it can be shown that a particle moving in the flame as described would adjust to within 10% of the gas velocity in 20 mm if it had a radius of 1×10^{-6} m or of 50 mm if the radius were 10×10^{-6} m. These distances are consistent both with the particles attaining the maximum rather than the mean velocity in the burner tubes and with the production of 10-mm long vapor trails during evaporation.

These remarks presume that the vaporized salt remains unmixed, but this neglects the effects of diffusion. Examination of the spike width at different heights in the flame (times) showed that the width increased linearly with time. Diffusion from a long cylinder is a complex problem, but since the probe is effectively stationary and the vapor trails aligned with the gas flow, only longitudinal diffusion must be considered, and the actual results of Figure 5 lead to an absolute diffusion coefficient of 3×10^{-4} m^2 sec^{-1} for the K$^+$ ion in a hydrogen flame at 2500 K. The measurements of Moseley (*13*) lead to values between 1.7 and 5×10^{-4} m^2 sec^{-1} depending on the temperature dependence assumed, but it is not certain that this is a proper comparison. The diffusion coefficient of neutral K would be expected to be larger than for the ion, but since ionization is itself a slow process, the possibility that diffusion occurs as neutral species cannot be ruled out.

Conclusion

Experimental studies on the ionization associated with particles in flames show that while such ionization occurs to a measurable extent, it is frequently associated with and even masked by ionization derived from gaseous material. The volatilization of such material can be a slow process and the dispersion of the vapor by diffusion even slower, so that each particle is marked by a meteor trail of ionized gas. Even when the particle has volatilized completely, such a trail may give the appearance of particulate matter. One result of this microscopic inhomogeneity is that estimates of the evaporation of droplets which assume a uniform distribution of vapor can be grossly in error.

Literature Cited

1. Hollander, T., Thesis, Utrecht, 1964.
2. Calcote, H. F., *Combust. & Flame* (1957) **1**, 385.
3. Page, F. M., Miller, E. R., *Combust. & Flame,* in press.
4. Page, F. M., *Phys. Chem. Fast React.* (1973) **1**, 161.
5. Pungor, E., "Flame Photometry Theory," p. 66, Van Nostrand, London, 1963.
6. Hayhurst, A. N., Kittelson, D. B., *Proc. Roy. Soc.* (1974) **A338**, 175.

7. Sugden, T. M., Thrush, B. A., *Nature* (1951) **168**, 703.
8. Smith, F. T., *Proc. Conf. Carbon, 3rd* (1959) 419.
9. Dimick, Soo, *Phys. Fluids* (1964) **7**, 1638.
10. Woolley, D. E., Thesis, Aston, 1968.
11. Egerton, A., Rudrakanchana, V., *Proc. Roy. Soc.* (1954) **A225**, 427.
12. Lapple, C. E., Shepherd, C. B., *Ind. Eng. Chem.* (1940) **32**, 605.
13. Moseley, J. T., Gatland, I. R., Martin, M. W., McDaniel, E. W., *Phys. Rev.* (1969) **A234**, 178.

RECEIVED December 29, 1976.

Formation of Large Hydrocarbon Ions in Sooting Flames

GILLES P. PRADO and JACK B. HOWARD

Department of Chemical Engineering, Massachusetts Institute of Technology, Cambridge, MA 02139

The concentrations of ionic species, large hydrocarbon molecules, and soot particles, and the mass distribution of ionic species larger than about 300 amu were measured previously in an acetylene/oxygen flame at 20 mm Hg using molecular beam sampling, laser absorption, electron microscopy, and a Langmuir probe. Assessment of the combined data indicates that the predominant ionic species are large hydrocarbons, probably polyaromatic, and young soot particles. The concentration of ions is large enough to support the view that ionic nucleation plays an important role in the formation of soot particles in flames. Possible mechanisms of ionization of large polynuclear aromatic hydrocarbons in flames are discussed.

There is evidence that ionic processes play a role in the formation of soot particles in flames, but it is not clear which step of soot formation—namely particle nucleation, surface growth, or agglomeration—is more affected by ions. The ions may serve as nuclei or precursors of nuclei, and charges residing on particles give rise to interparticle electrostatic forces that are significant in the agglomeration process. The identification of flame ions and the measurement of their concentration have been limited usually to lean flames (1, 2). Experiments concerned with ionic effects in sooting flames have focused on the effects of ionic additives (3, 4) or of electric fields (5, 6) on the amount of soot produced but have not provided direct information on the ionic species formed in unperturbed sooting flames.

In order to understand the effect of ions on soot formation, it is important to understand both the nature of the ionic species present in sooting flames and the mechanism of their formation. Some progress in that direction has been made in recent experiments at the Massachusetts Institute of Technology. This chapter describes the various observations of large ions in sooting flames and discusses the different mechanisms believed to be responsible for the formation of these species.

Experimental

The experiments were performed by Wersborg (7), Prier (8), Yeung (9), and Adams (10). The apparatus used by Wersborg, Yeung, and Adams has been described elsewhere (11, 12). Flat or one-dimensional flames of premixed acetylene and oxygen were burned in a chamber at 20 mm Hg on a 7-cm diameter water-cooled burner consisting of a copper plate having uniformly distributed small holes. The flames were sampled at different positions along the centerline using a molecular beam instrument which provided quench times of about 1 μsec. Ionic species were analyzed using a simple mass spectrometer installed in the beam system giving low but adequate mass resolution in the range $\sim$ 250 to a few thousand amu. The number concentration of charged species was determined by measuring the electric current delivered to a Faraday cage intercepting the beam which contains a known volume flow rate of flame sample. The mass distribution of charged species was determined by the deflection of increasingly large species from the beam using increasingly large electrical deflection voltages. The concentration and size distribution of all particles (charged and neutral) larger than about 15 Å diameter were determined by high-resolution electron microscopy of beam deposits collected directly on microscope grids.

The molecular beam filtered out electrons and small ions since no electrostatic focusing was used. Applied deflection voltage less than about 0.1 V had no effect on the current arriving at the Faraday cage, which indicates that ions smaller than about 300 $\pm$ 100 amu were not in the beam.

In order to measure the total ion concentration (i.e., small and large ions) in sooting flames, Prier used a burner designed to duplicate the flames used in the beam-sampling apparatus and a Langmuir probe consisting of two parallel platinum wires which could be positioned horizontally at any distance above the burner. Measurement of the electrical current between the two wires as a function of the applied potential difference provided information about the plasma conditions at the probe site, from which data the ion concentrations were calculated. The soundness of the apparatus and the accuracy of the measurements were tested under relatively well known non-sooting conditions by duplicating an experiment of Calcote (13) using a propane–air flame at 33 mm Hg, a fuel equivalence ratio of 0.875, and a total flow rate of 122 cm^3/sec. Assuming a temperature of 2080 K and the predominant ion to be H_3O^+, a value of 0.45×10^9 cm^{-3} was obtained for ion concentration. With the same conditions and assumptions, Calcote obtained a concentration of 0.6×10^9 cm^{-3}. The agreement is excellent considering the uncertainty involved.

The total concentration of heavy hydrocarbon molecules as a function of height above the burner in sooting flames was determined by measuring the attenuation of a laser beam passed horizontally through the flame (*14, 15*). The required soot absorption coefficients at the different flame positions were measured in the same optical system by replacing the flame with soot samples deposited on glass slides in the molecular beam sampling instrument and characterized by electron microscopy. The concentration of large hydrocarbon species was estimated by subtracting from the total signal the absorption from soot and by assuming the absorption coefficient of large hydrocarbon species to be the same as that measured for young soot particles.

Results

Although all of the techniques described above were applied in acetylene–oxygen flames at 20 mm Hg, experimental difficulties prevented their application to the same conditions of fuel equivalence (Φ) and cold gas velocity (v).

Selected results on number concentrations are presented on Figure 1. Wersborg's results were obtained at $\Phi = 3.0$ and $v = 50$ cm/sec. The soot particles (curve a) refer to particles larger than about 15 Å diameter, measured by electron microscopy. The charged fraction of particles (curve b) was determined by measuring particle number with and without an electric field applied across the beam to remove all charged particles. Measurement error was estimated to be below 20%.

Yeung's results (curve c) were obtained by measuring the electric current delivered to a Faraday cage, as described above. The fuel equivalence ratio was also 3.0, but the cold gas velocity was 38 cm/sec. An increase in cold gas velocity decreases the ion concentration (*9, 10*), apparently by as much as a factor of 10 between 38 and 50 cm/sec. We choose to present Yeung's results in the absence of more precise data, but the influence of v has to be kept in mind when comparing curves a and c.

Similar remarks apply to Prier's data (curves d and e), where v is 50 cm/sec, but Φ is 3.5 instead of 3.0. Decreasing the fuel equivalence ratio also decreases the ionic concentration, as shown by Yeung (*9*) and Adams (*10*) for Φ smaller than 3.0. The effect of Φ between 3.0 and 3.5 is not known, but the trend in the data indicates that it may be a factor of 10.

The calculation in Prier's work of ionic concentrations using measurements of saturation currents required knowledge of the mobility and size of the predominant ionic species. Although this information was unavailable at the time of Prier's thesis, the shape of the concentration profiles with height above the burner and the order of magnitude of the concentrations led Prier to conclude that the predominant charged species were particles around 40 Å in diameter. Later, Wersborg measured size

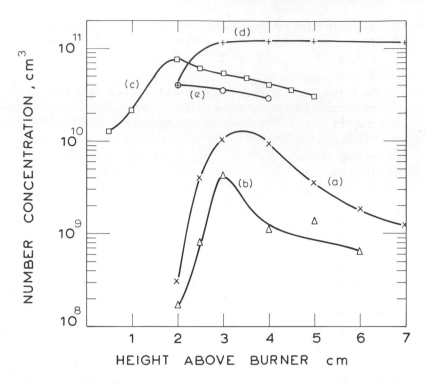

Figure 1. Number concentrations of species in sooting acetylene/ oxygen flames as a function of height above the burner for different fuel equivalence ratios (ϕ) and cold gas velocities (v). Pressure = 20 mm Hg; points = experimental data; curves = estimated trends. (a) Total soot particles; (b) charged soot particles: ϕ = 3.0, v = 50 cm/ sec, Wersborg (7); (c) large positive ions: ϕ = 3.0, v = 38 cm/ sec, Yeung (9); (d) and (e) total positive ions, different assumed species sizes (see text), ϕ = 3.5, v = 50 cm/ sec, Prier (8).

distributions of soot particles and Adams and Yeung measured charge/ mass ratio distributions of charged species. Their results are reported in Table I as a function of height above the burner. Considering the different experimental conditions, the results of Yeung and Adams are in good agreement.

Also presented in Table I are temperatures of the flame as measured by others(*16, 17*) and the saturation currents measured by Prier. With the assumption that no collisions between positive ions and neutral gas molecues occur within the ionic sheaths surrounding the probe wires (valid here because the mean free path of the species considered is larger than the thickness of the sheath), the relation between the saturation current and the ionic concentration is:

$$n_+ = 4.01 \times 10^{16} \, (r^3 \rho / T)^{1/2} \, i_s$$

where i_s = positive ion saturation current, (A); r = particle radius (A); ρ = density, assumed to be 1.8 g/cm³ for the carbonaceous material of interest here; T = temperature (K); and n_+ = number concentration (cm⁻³). The assumption that soot particles are the main ionic species leads to the curve d in Figure 1, computed by using Wersborg's particle size. Using Adam's ion sizes, curve e is obtained. The data of Yeung were not used in these calculations, the experimental conditions being too different. Curve e is probably closer to the actual concentrations than curve d because the large soot aggregates tend to be neutralized by electron recombination and because ions smaller than 15-Å diameter certainly exist but have escaped observation by electron microscopy. Within the limitations cited above it appears that:

(1) Ionic concentrations measured with the different techniques are in fairly good agreement.

(2) There are enough ions in these flames to support much interest in the ionic nucleation of soot particles.

(3) The main ionic species in sooting flames are heavy hydrocarbon molecules and young soot particles, with a mass in the range ~ 300 to several thousand amu.

In addition, Figure 2 presents a comparison between the volume of heavy hydrocarbon molecules (neutral and ionized) measured by light absorption (14, 15) and the volume of soot particles measured by molecular beam sampling and electron microscopy (7, 11). These data show that enough heavy molecules exist to account for soot formation, thereby supporting the view that these molecules are probably intermediates of soot.

Table I. Temperature, Saturation Current (*I*), and Diameter of the Most Abundant Observed Ions (*D_A*) as a Function of Height above Burner for Different Fuel Equivalence Ratios (*φ*) and Cold Gas Velocities (*v*)

Height above Burner (cm)	Temperature (K) ($\phi = 3.5$, v = 50 cm/sec) (16, 17)	$I \times 10^6$ (A) ($\phi = 3.5$, v = 50 cm/sec) (8)	D_A of Soot Particles (Å) ($\phi = 3.0$, v = 50 cm/sec) (7)	D_A of Ionic Species (Å) ($\phi = 2.25$, v = 31.3 cm/sec) (9)	D_A of Ionic Species (Å) ($\phi = 3.0$, v = 50 cm/sec) (10)
1.25	2107	.48	—	13	—
1.50	2098	.65	—	18	—
1.75	2085	.42	—	20	—
2	2072	.38	40	22	40
3	2036	.27	100	—	45
4	2000	.19	130	—	50
5	1964	.13	170	—	—
7	1893	.066	250	—	—

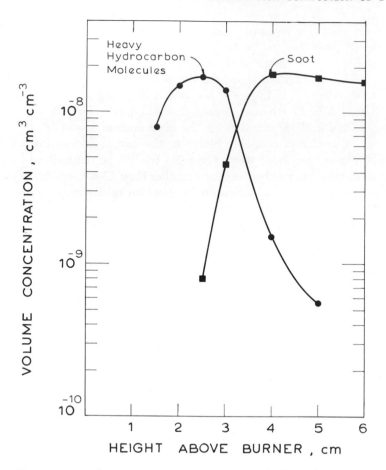

Figure 2. Volume concentration of heavy hydrocarbon molecules and soot at different heights above the burner in an acetylene/oxygen flame. Pressure = 20 mm Hg; fuel equivalence ratio = 3.0; cold gas velocity = 50 cm/sec.

The foregoing observations lead us to support a mechanism of ionic nucleation whereby large hydrocarbon ions are formed which can then agglomerate large neutral hydrocarbons to form soot particles by surface growth. The chemical nature of the large hydrocarbon ions and the mechanism of their formation are discussed below.

Tentative Mechanism

The chemical nature of the large hydrocarbon molecules in sooting flame is not well established, but there is much evidence that they are polyaromatic hydrocarbons and their alkyl derivatives. Bonne, Homann, and Wagner (*16*) identified polyacetylenes up to a mass of 146 amu, but not higher. In the mass range 100–350 amu, several authors identified polyaromatic hydrocarbons and their alkyl derivatives (*17, 18, 19, 20*). Experimental difficulties have prevented the identification of higher mass compounds, but polyaromatic hydrocarbons are probably the only species with masses in the range of the ionic species described above. Consequently, we assume in the following discussions that the main ionic species in the soot formation zone of flames are polynuclear aromatic hydrocarbons and young soot particles.

Two general mechanisms are usually advanced to explain ionization of molecules in flames: direct ionization by thermoionization, photoionization, or chemiionization and indirect ionization by charge transfer with other ions. The assessment of both mechanisms requires knowledge of the ionization potential of molecules. In the following discussion, computations developed in the Appendix are used to estimate approximate ionization potentials of polynuclear aromatic hydrocarbons.

Thermoionization. A fraction of the large hydrocarbon molecules and soot particles, both represented by R, can be ionized directly via the reaction:

$$R \rightarrow R^+ + e^-$$

If we assume in an orientation calculation with the Saha equation (*21*) that all the species have the same ionization potential (*I*) and a temperature of 2000 K, the fraction of species charged is given by:

$$n_+/n = (3.7 \times 10^{26}/n_e) \exp(-6.44I)$$

where n_e is the electron concentration in cm^{-3}, and I is in eV. Values of n_+/n thus calculated are shown in Table II for typical electron concentrations in the flames of present interest and ionization potentials estimated as described in the Appendix.

Table II. Fraction of Ionized Particles as a Function of Their Diameter (D_A) and Electron Concentration (n_e), Computed by Using the Saha Equation

D_A (Å)	n_e (cm^{-3})		
	10^9	10^{10}	10^{11}
10^a	1.8×10^{-5}	1.8×10^{-6}	1.8×10^{-7}
10^b	3.4×10^{-7}	3.4×10^{-8}	3.4×10^{-9}
20^a	1.9×10^{-3}	1.9×10^{-4}	1.9×10^{-5}
30^a	9×10^{-3}	9×10^{-4}	9×10^{-5}

[a] Spherical particles.
[b] Planar molecules.

Using the maximum volume concentration of heavy hydrocarbon molecules from Figure 2 and assuming that all the particles have the same size, one can compute the particle number concentration (n) corresponding to the different particle sizes in Table II. The values obtained range from $n = 3.5 \times 10^{13}/cm^3$ for 10-Å planar molecules to $n = 1.2 \times 10^{12}/cm^3$ for 30-Å spherical particles. Use of these values and the extents of ionization shown in Table II predict ionic concentrations that are much smaller than the experimental values shown in Figure 1, the discrepancy being larger for the smaller assumed sizes and reaching several factors of 10 in the case of 10-Å planar molecules. The largest predicted ionic concentration is $10^{10}/cm^3$, for the case of 30-Å spherical particles and an electron concentration of $10^9/cm^3$ (the local electron concentration can be lower than positive ion concentration because of higher electron mobility). This concentration is not unrealistic compared with the experimental values of $10^{10} - 10^{11}/cm^3$, but the size of the ions present when soot particles first begin to form is much closer to 10 Å than 30 Å (see Yeung's data in Table I). Therefore, thermoionization plays no more than a minor role in the primary zone of soot formation.

Photoionization. Large polyaromatic molecules are very easily oxidized when adsorbed on a solid (22). Such behavior is observed on thin layer chromatography plates, and it also explains the destruction of polycyclic aromatic hydrocarbons adsorbed on particulates in the atmosphere. The oxidation may occur via the formation of an intermediate radical cation, probably under natural or ultraviolet light. The radical cations are unstable and react rapidly with nucleophiles, including unoxidized hydrocarbon, or with oxygen. As an example, dimer formation is observed during the anodic oxidation of benzo[a]pyrene. Such radical cation can be formed presumably also in flames, especially when large polyaromatic molecules are adsorbed on carbon particles. The radical character of young soot particles is well known (17) and can be enhanced by exposure to light (23).

Although it is not yet possible to assess the effect of photoionization in flames, this mechanism may well have an important contribution to ion formation, especially toward the end of the process when polyaromatic hydrocarbons are adsorbed on the surface of soot particles.

Chemi-ionization. Chemi-ionization is the main and probably the only mechanism responsible for the formation of primary ions in fuel-lean flames. This mechanism, first proposed by Calcote (13), has received much attention, and there is general agreement that CHO^+ is the only primary chemi-ion formed in lean flames (see below).

In the case of fuel-rich flames, we found no mention in the literature of reactions involving large hydrocarbon molecules and releasing enough energy to expel an electron. Although such reactions may exist, no discussion on this mechanism will be attempted here.

Ionization by Charge Transfer. Many ion–molecule reactions between small species (< 50 amu) occur in flames (24). As mentioned above, the only primary ion in fuel-lean flames (non-sooting) is CHO^+, formed by chemi-ionization ($13, 24, 25$).

$$CH + O \rightarrow CHO^+ + e^- \qquad \Delta H = 8 \pm 20 \text{ kJ/mol } (26)$$

The more abundant ions are H_3O^+, $C_3H_3^+$, and $C_2H_3O^+$, usually believed to be formed by ion molecule reactions such as:

$$CHO^+ + H_2O \rightarrow H_3O^+ + CO$$

In sooting flames there are indications that charge transfer may occur between small ions such as $C_3H_3^+$ and H_3O^+ and large polynuclear aromatic hydrocarbons. Indeed, the ionization potential of these large molecules is small enough so that the charge transfer is thermodynamically favorable. For example, the ionization potential for the production of $C_3H_3^+$ is 11.3 eV, higher than that of any aromatic compound, and for H_3O^+ it is 6.3 eV, larger than that of any polyaromatic of mass larger than about 600 amu. Furthermore, increasing the fuel equivalence ratio beyond $\Phi = 1$ decreases the concentration of small ions in a non-sooting flame (Figure 3) and increases the concentration of large ions in a sooting flame (Figure 4). This behavior supports the concept that small ions transfer their charge to large molecules, the concentration of the latter increasing with increasing fuel equivalence ratio.

Critical assessment of the foregoing concept would require precise identification of small ions in the oxidation zone of a sooting flame. Such information would clarify the nature of the primary ions. Indeed, the belief that CHO^+ is the only primary ion formed in sooting systems was recently questioned by Abrahamson and Kennedy (27). These authors observed charge carriers to be associated with soot formation during the pyrolysis of hydrocarbons in absence of oxygen, which indicates that ion formation in sooting flames is not limited to mechanisms involving CHO^+. Possible alternatives include direct ionization of large molecules as described above or charge transfer from chemi-ionization products other than CHO^+. For example, $C_3H_3^+$ formed by the reaction:

$$CH + C_2H_2 \rightarrow C_3H_3^+ + e^-$$

may well be important in sooting flames, especially those of acetylene, even though this reaction has been virtually eliminated as being important in lean flames (28).

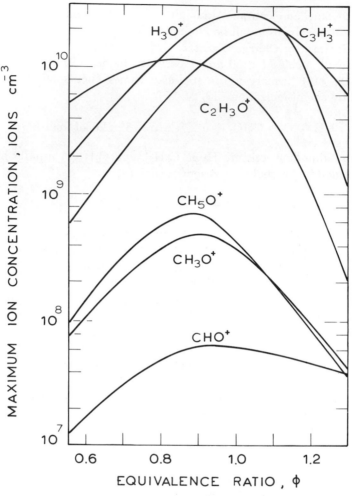

10th International Symposium on Combustion

*Figure 3.   Effect of mixture composition on positive ion con-
centrations in a non-sooting flame (26).   Acetylene/oxygen;
pressure = 4 mm Hg.*

Finally, charge transfer reactions have been advanced also to explain
the effect of metallic additives on soot emission from flames. The ability
of metallic additives either to diminish or to enhance the soot formation
in flames has been well known for many years (3). Their role is usually
linked either to modification of the production rate of hydrocarbon ions
(4) or to alteration of the concentration of OH radicals (3). Although
there is still some disagreement about experimental results as well as their
interpretation, it now appears (29) that the soot-promoting effect is
linked to ions resulting from metallic additives through reactions such as:

$$M^+ + R \rightarrow M + R^+$$

and its reverse, followed by ionic nucleation. Here M and R represent a metal atom and a large hydrocarbon molecule, respectively. For example, in a study of the effects of additives in an acetylene/oxygen diffusion flame, Bulewicz (4) observed that adding a small amount of cesium has a pro-sooting effect while adding a larger amount has an anti-sooting effect, the transition occurring smoothly for an ionic concentration of about 2×10^{10} ions/cm^3. The number of soot particles emitted first increases and then reaches a maximum at the additive ion null-point

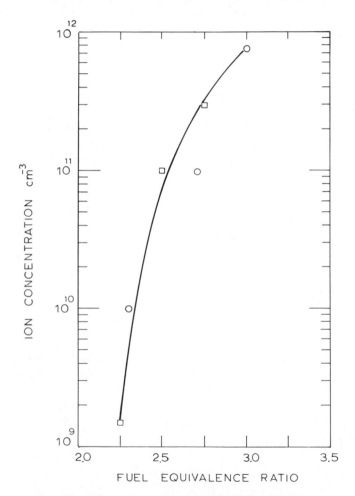

Figure 4.   Effect of mixture composition on positive ion concentration in a sooting flame.   Acetylene/oxygen; pressure = 20 mm Hg; cold gas velocity = 31.3 cm/sec; (□) Ref. 9; (○) Ref. 10.

concentration. This increase in number of soot particles in the pro-sooting regime, which was observed earlier by Weinberg (30), is evidence that ions affect the nucleation process. There is clearly much potential for future work along these lines.

Summary

This chapter critically assesses data obtained previously from an acetylene/oxygen flame at 20 mm Hg. The measurements, performed along the centerline of a one-dimensional flame, included the total concentration of ionic species, large hydrocarbon molecules, and soot particles and the mass distribution of ionic species larger than about 300 amu. Charge concentration of species larger than about 300 amu was determined by measuring the electric current delivered to a Faraday cage in the detection chamber of a staged molecular beam flame sampling instrument, and charge/mass ratio distributions were measured by the incremental electrical filtration of charged species from the beam. The concentration profiles of soot were obtained by electron microscope analysis of beam deposits while concentration profiles of large hydrocarbons were estimated by subtracting the contribution of soot particles from the attenuation signal of a laser beam passed through the flame. Total charge concentration was measured with a Langmuir probe. The data indicate that the predominant ionic species present in the sooting region of the flame are large hydrocarbon molecules, probably polyaromatic in nature, and young soot particles. The concentration of ions is high enough to justify the continuing interest in ionic nucleation of soot particles in flames. Possible mechanisms of ionization of large polynuclear aromatic hydrocarbons are discussed.

Appendix

Calculation of Ionization Potentials. It is possible to perform an approximate calculation of the ionization potentials of young soot particles and aromatic hydrocarbon molecules (31, 32, 33). Assuming the species to be an electrically conducting sphere or circular disk of radius r and capacitance C, the first ionization potential, i.e., the work required to remove one electron, is:

$$I = V + e^2/2C$$

where V is the work function, e is the charge of an electron, and C is $4\pi\epsilon_0 r$ for spheres and $8\epsilon_0 r$ for disks, ϵ_0 being the dielectric constant of space ($= 8.85 \times 10^{12}$ F/m).

Table A-I. Ionization Potential of Polyaromatic Molecules

	M(amu.)	I(eV)[a]	I(eV)[b]	I(eV)[c]	I(eV)[d]
Naphthalene	128	10.58	—	8.25	8.26
Anthracene	178	9.44	—	7.30	7.55
Pyrene	202	8.76	—	7.68	7.55
Coronene	300	7.70	6.19	6.50	7.60
Ovalene	398	7.16	6.07	5.64	—

[a] Capacitive energy model—planar molecules.
[b] Capacitive energy model—spherical molecules.
[c] Electronic levels model.
[d] Experimental values from Ref. *31*.

As a first approximation, we assume V to be the work function of graphite (4.39 eV). For spherical particles, r is calculated from the molecular weight M assuming a density of 1.8 g/cm^3. Thus:

$$r = 0.583 \ (M)^{1/3} \ \text{Å}$$

where M is in g/mol.

For planar molecules, r is approximated as the radius of a disk having the same area as the molecule. This area is computed from the number of hexagonal rings in the structure, the area of one ring being about 5.24 Å^2.

The ionization potential can also be calculated by considering the electronic levels (*34*). Thus

$$I = (39 + 7n)R/k^2$$

where R is the Rydberg constant (13.53 eV), n is the number of supplementary sextets (after the first), and k is the order number which is 7.5 for benzene and increases by 0.5 for each additional ring. Comparisons between values calculated using these different methods and experimental values when known are reported in Table A-1. The electronic model is apparently more accurate for masses smaller than 202 (pyrene). For larger masses, the capacitive energy model with planar molecules seems accurate (cf. coronene) even though particles are probably not conductors. In the absence of better information, we therefore used the capacitive energy model with spherical particles larger than 800 amu.

Literature Cited

1. Calcote, H. F., "Ions in Flames," in *Ion-Mol. React.* (1972) **2**, 673.
2. Miller, W. J., *Oxid. Combust. Rev.* (1968) **3**, 97.
3. Cotton, D. H., Friswell, N. J., Jenkins, D. R., *Combust. Flame* (1971) **17**, 87.
4. Bulewicz, E. M., Evans, D. G., Padley, P. J., *Proc. Symp. (Int.) Combust.*, *15th* (1975) 1461.

5. Lawton, J., Weinberg, F. J., "Electrical Aspects of Combustion," Oxford, 1969.
6. Heinsohn, R. J., Becker, P. M., "Effects of Electric Fields in Flames," in *Combust. Technol. Some Mod. Dev.* (1974).
7. Wersborg, B. L., "Physical Mechanisms of Carbon Black Formation in Flames," Ph.D. Thesis, M.I.T., 1972.
8. Prier, L., "Measurement of Electron Concentrations in Sooting Hydrocarbons Flames," B.S. Thesis, M.I.T., 1972.
9. Yeung, A. C., "Concentration and Mass Distribution of Charged Species in Sooting," Flames," M.S. Thesis, M.I.T., 1973.
10. Adams, D. E., "Measurement Technique for Charged Soot Particles in Flames," Chemical Engineer's Degree, M.I.T., 1976.
11. Wersborg, B. L., Howard, J. B., Williams, G. C., *Proc. Symp. (Int.) Combust., 14th* (1975) 929.
12. Wersborg, B. L., Yeung, A. C., Howard, J. B., *Proc. Symp. (Int.) Combust., 15th* (1975) 1439.
13. Calcote, H. F., *Proc. Symp. (Int.) Combust., 8th* (1962) 180.
14. Fox, L. K., "Optical Determination of Soot Concentration in a Low Pressure Oxyacetylene Flame," M.S. thesis, M.I.T., 1972.
15. Wersborg, B. L., Fox, L. K., Howard, J. B., *Combust. Flame* (1975) **24**, 1.
16. Homann, K. H., Wagner, H. G., *Ber. Bunseges. phys. chem.* (1965) **69**, 20.
17. Bonne, U., Homann, K. H., Wagner, H. G., *Proc. Symp. (Int.) Combust., 10th* (1965) 503.
18. Tompkins, E. E., Long, R., *Proc. Symp. (Int.) Combust., 12th* (1969) 625.
19. D'Alessio, A., DiLorenzo, A., Sarofim, A. F., Beretta, F., Masi, S., Venitozzi, C., *Proc. Symp. (Int.) Combust., 15th* (1975) 1427.
20. Prado, G. P., Lee, M. L., Hites, R. A., Hoult, D., Howard, J. B., *Proc. Symp. (Int.) Combust., 16th* (1977) 649.
21. Saha, M. N., Saha, H. K., "A Treatise of Modern Physics," Vol. I, The Indian Press, 1934.
22. "Particulate Polycyclic Organic Matter," Committee on Biologic Effects of Atmospheric Pollutants, Division of Medical Sciences, National Research Council, National Academy of Sciences, Washington, D.C., 1972.
23. Donnet, J. B., Morawski, J. C., personal communication, 1971.
24. Miller, W. J., *Proc. Symp. (Int.) Combust., 14th* (1973) 307.
25. Zallen, D. M., Hirleman, E. D., Wittig, S. L. K., *Proc. Symp. (Int.) Combust., 15th* (1975) 1013.
26. Calcote, H. F., Kurzius, S. C., Miller, W. J., *Proc. Symp. (Int.) Combust., 10th* (1965) 605.
27. Abrahamson, J., Kennedy, E. R., "Twelfth Biennial Conference on Carbon," p. 169, The American Carbon Society, 1975.
28. Miller, W. J., *Proc. Symp. (Int.) Combust., 11th* (1967) 311.
29. Feugier, A., "European Symposium of Combustion Institute," p. 406, Academic, 1973.
30. Place, E. R., Weinberg, F. J., *Proc. Symp. (Int.) Combust., 11th* (1967) 245.
31. Smith, F. T., *J. Chem. Phys.* (1967) **34**, 793.
32. Soo, S. L., Dimick, R. C., *Proc. Symp. (Int.) Combust., 10th* (1965) 699.
33. Howard, J. B., *Proc. Symp. (Int.) Combust., 12th* (1969) 877.
34. Clar, E., "Polycyclic Hydrocarbons," p. 105, Acadlemic, 1964.

RECEIVED December 29, 1976.

11

Radical and Chemi-ion Precursors:
Electric Field Effects in Soot Nucleation

S. L. K. WITTIG[1] and T. W. LESTER[2]

Combustion Laboratory, School of Mechanical Engineering,
Purdue University, West Lafayette, IN 47907

*Radical and chemi-ion participation in the earliest stages
of soot formation has been studied in the reflected shock
region of a conventional shock tube. Electrostatic probes,
laser line extinction, and absorption spectroscopy were used
to follow the time history of positive ions, nascent soot particles, and radical species. Fundamentally different behavior
in fuel-rich vs. fuel-lean combustion was noted for positive
ions, a result that can be plausibly explained by a different
charge exchange mechanism in fuel-rich conditions. In a
separate series of experiments electric fields were used to
perturb the space charge neutrality and nucleation. The
fields had a varying influence on nucleation, inhibiting it in
mixtures just above the carbon forming-limit but augment-
ing it in very rich cases.*

The nucleation of soot from gaseous hydrocarbon combustion has been
of longstanding fundamental interest. Recently, pioneering experiments by Wersborg et al. (*1, 2*), Weinberg et al. (*3, 4*), and Howard (*5*)
have emphasized the role of ionic species in the formation route of soot.
While suggesting ionic participation from the very earliest stages of soot
formation, flame experimentation has failed to demonstrate conclusively
the actual role of chemi-ions. The spatial resolution of kinetic steps has
been hampered by transport phenomena resulting from the concentration
and thermal gradients in the vicinity of the flame and post-flame region.
In fact, previous positive ion measurements may not present accurately

[1] Present address: Institut für Thermische Strömungsmaschinen, Universität
Karlsruhe, Kaiserstrasse 12, D-7500 Karlsruhe 1, West Germany.
[2] Present address: Department of Nuclear Engineering, Kansas State University,
Manhattan, KS 66506.

the actual concentration of ionic species because of the substantial bulk velocity and diffusion effects which have been largely ignored in the reduction of Langmuir probe data.

To overcome this difficulty and to resolve the sequential processes leading to the first soot particles, the reflected shock region of a conventional shock tube has been used. The core region of the reflected shock provides a largely homogeneous reaction zone where concentration and thermal gradients are reduced greatly compared with atmospheric flames, for example.

This chapter details the experimental apparatus and technique as well as the experimental results, discusses the observations obtained here in light of previous flame data, and discusses the initial steps of a kinetic mechanism.

Experimental Apparatus

The shock tube and basic peripheral instrumentation have been described previously by Lester et al. (6, 7). In their experiment a simple attenuation technique was used to monitor the soot formation. The primary light source, shown in Figure 1, was a Spectra-Physics model 120,

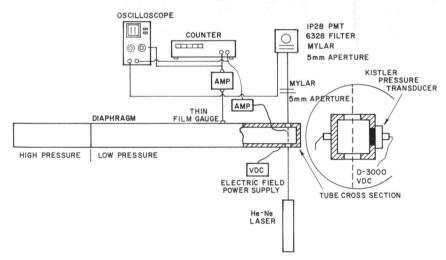

Figure 1. Experimental apparatus

5-mW He–Ne laser. The laser exhibits a 0.8-mm beam width and less than 1-mrad divergence. To eliminate unwanted emission from the hot test gas and incandescent soot particles, two 5-mm apertures were used in conjunction with Mylar diffusing screens. As a final guard against emission interference, the beam was passed through a laser-line filter (6328 Å wavelength, 100-Å half-width) mounted in front of the magnetic μ-shielded photomultiplier tube (RCA 1P-28). The entire photomultiplier assembly, including laser-line filter, was enclosed in a light-proof box. The selectivity of the system was such that no emission signal was recorded when the gas mixture was shock-heated. Linearity of response was verified by sequential testing with a series of neutral density filters. The sensitivity of the apparatus allowed detection of soot particles down to a volume fraction of about 5×10^{-9}.

Positive ion data were monitored with a cylindrical Langmuir probe similar in design to that used in previous experiments (7). The probe circuitry was similar to that of Hand and Kistiakowsky (8). The probe was biased at voltages from -40 V to $+40$ V, and the resulting characteristic curve data were reduced following Calcote (9), in which the positive-ion saturation curve is extrapolated to the plasma potential to minimize the uncertainty in the collection area.

The electric field runs were conducted in a companion shock tube with ports for inserting a large cylindrical electrode (*see* Figure 1). The electrode was a 5.4-cm diameter copper electrode mounted in a Plexiglas insulator. The assembly was installed flush with the bottom wall. The electrode/tube apparatus operated much like an ionization chamber. A D.C. power supply provided continuous output from 100 to 4000 V, giving a field strength of up to 500 V/cm.

The pre-mixed methane/argon mixture consisted of 1.2% methane and 98.8% argon, both being 99.99% Matheson research grade. Argon/oxygen and zero air were used as oxidants. Prior to each run the driven section was evacuated with an oil diffusion pump. Tests were conducted in shock-heated argon to determine possible interference from sodium impurities; however, no ionization was detected.

Experimental Results

Comprehensive data were collected on positive ion and soot behavior over a temperature range of 1500° to 3000°F and at reactant partial pressures of approximately 1 atm. The present experimental conditions more nearly match the pressures and temperatures found in conventional combustion systems than do the artificially low pressures usually needed in flame experimentation with comparable resolution.

Typical pressure history and electrostatic probe response for rich methane combustion are shown in Figure 2. For comparison, Figure 3 displays a chemi-ion trace obtained at an equivalence ratio of .476. It is immediately apparent that the profiles differ, especially after the peak concentration is obtained. While the probe current in the fuel-lean system rapidly decays (compare also Refs. 6, 7), the ion current in the fuel-rich case decreases only slightly to an equilibrium value. It is possible that this intrinsic difference may result from ion–molecule reactions competing with recombination, thereby causing the total charge in the fuel-rich system to extend longer into the post-ignition region. Evidence for this is discussed in the next section.

The effects of the determining combustion parameters on the positive ion yield are compared in Table I. It is important that the total yield

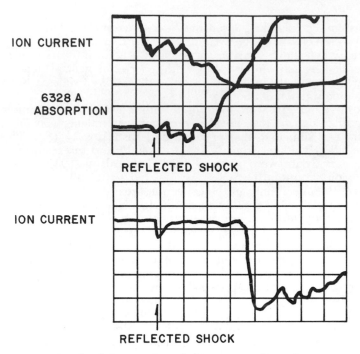

Figure 2. (top) Soot and ion behavior in rich methane combustion. $\phi = 2.86$; $T_5 = 2836$ °K; $P_{5,REAC} = .824$ atm; sweep $= .5$ msec/cm.

Figure 3. (bottom) Chemi-ion behavior in lean methane combustion. $\phi = 0.476$; $T_5 = 2187$ °K; $P_{5,REACT} = 3.03$ atm; sweep $= .5$ msec/cm.

decreases by an order of magnitude from an equivalence ratio of .50 to 2.86. The ion yield at a very rich condition, $\phi = 2.95$, but at a much hotter temperature is also documented in the table. The yield is approximately the same as it is for a leaner system at a lower temperature. Although

Table I. Positive Ion Concentration Data

Equivalence Ratio	T(°K)	P(atm)	Positive Ion Concentration $n_+(cm^{-3})$	Ion Yield per Fuel Molecule $\left[\dfrac{n_+}{CH_4}\right] \times 10^4$
.50	2175	5.1	1.58×10^{13}	.48
.50	2445	6.0	2.08×10^{13}	.60
1.07	2694	6.0	3.63×10^{13}	1.41
1.62	2458	4.0	3.47×10^{12}	.11
2.86	2388	7.2	3.26×10^{12}	.05
2.95	2733	4.3	1.50×10^{13}	.26

the ion yield is comparable, the qualitative behavior remains quite different.

To determine the influence of positive charges on the early soot formation route, laser line absorption profiles were recorded in a separate series of runs without and with electric fields applied across the reflected shock region. The electric field in these experiments, less than 500 V/cm, is not sufficient to manipulate positive ions or charged particles, and thus it differs fundamentally from the flame work of Weinberg et al. (*3, 4*). The reaction region is, however, analogous to a parallel plate ionization chamber in which electrons are removed from the gas before recombination reactions with positive ions occur. The saturation current technique, as described by Lawton and Weinberg (*10*), has been used in shock tubes to study chemi-ionization by Matsuda and Gutman (*11*) and by Bowser and Weinberg (*12*). The first authors established that a field of 25 V/cm was sufficient to cause saturation to a 3.4-cm diameter electrode while the second experimenters observed that a field of 200 V/cm was

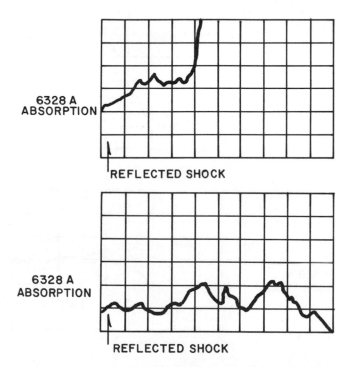

Figure 4(a)   (top) Soot behavior in slightly rich methane combustion with no applied field. $\phi = 1.82$; $T_5 = 2780$ °K; $P_{5,REAC} = 1.13$ atm; sweep $= 1$ msec/cm. (b) (bottom) Soot behavior in slightly rich methane combustion with an applied field of 300 VDC. $\phi = 1.82$; $T_5 = 2535$ °K; $P_{5,REAC} = 1.03$ atm; sweep $= 1$ msec/cm.

necessary to cause saturation current to a cylindrical probe. The important feature of the method as used in this experiment is that since the field can destroy only the space charge neutrality by removing electrons, any subsequent effect on soot formation must come about through positive ion involvement in the earliest nucleation steps.

Depending on the stoichiometry, the field either inhibited or enhanced the soot formation. Figure 4 compares a run with no field with a run with a moderate field of 300 V. The stoichiometry, $\phi = 1.82$, is just above the carbon-forming limit in this experimental apparatus. The applied field extinguishes completely the soot formation from its inception. The tests were extended subsequently to very rich conditions, $\phi = 2.86$. Figure 5 compares runs made with and without the applied field. The field effect is dramatically different. Instead of diminishing the soot yield, the field actually has augmented the soot formation.

Consequently, a qualitative difference was observed in chem-ion behavior between fuel-lean and fuel-rich combustion. The ion yield

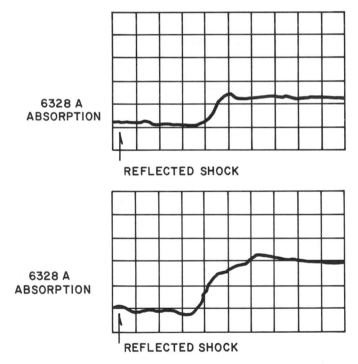

Figure 5(a) (top)   Soot behavior in very rich methane combustion with no applied field. $\phi = 2.86$; $T_5 = 2625$ °K; $P_{5,REAC} = .93$ atm; sweep $= 1$ msec/cm. (b) (bottom) Soot behavior in very rich methane combustion with an applied field of 300 VDC. $\phi = 2.86$; $T_5 = 2611$ °K; $P_{5,REAC} = .927$ atm; sweep $= 1$ msec/cm.

per fuel molecule was also an order of magnitude less in the rich system than in lean combustion under comparable conditions. Finally, when moderate electric fields were applied across the reflected shock region in a separate series of runs, soot formation was diminished in slightly sooting combustion, but it was augmented in very fuel-rich systems.

Discussion

Both the substantial magnitude of positive-ion formation in the reaction process and the effective control of soot formation through electric field application imply the intimate involvement of chemi-ions in the early soot formation mechanism, particularly in slightly rich combustion.

The total positive-ion yield at slightly rich and very rich conditions agrees with earlier work by Lester et al. (7) and by Zallen et al. (13). Microwave and electrostatic probe observations by Schneider and Grönig (14) in noble gases have yielded self-consistent ion and electron concentrations and give confidence that the electrostatic probe results are in fact sufficiently accurate in the present context. In a very recent study by Bowser and Weinberg (12), ion concentrations of 10^{11}–10^{15} ions/cm^3 were recorded in shock-initiated ethylene pyrolysis. From these shock tube studies, it seems certain, therefore, that sufficient positive charge carriers are produced through chemi-ionization in rich combustion to permit a substantial contribution of charged species in soot nucleation.

Homann (15) has argued that direct ion involvement is improbable in very rich combustion because the acetylenic species are present in substantially greater quantity. It has been proposed that these hydrocarbons build through polyacetylenic species to form embryonic soot. We believe that in very rich combustion this may be the primary route, but it appears to be only one route at lower stoichiometry.

The extensive flame work recently reported by Wersborg et al. (1, 16, 17) has extended Homann's earlier efforts to include charge effects. The dual-nucleation path suggested by Wersborg et al. indicates that the dominant nucleation path depends critically on the relative concentration of chem-ions and large gas-phase hydrocarbons. Under extremely rich conditions, such as studied by Homann, a neutral–neutral path would be expected to dominate because of the larger relative concentration of gas-phase hydrocarbons compared with chem-ions. Under these conditions, a relatively small fraction of the large hydrocarbons would participate in charge exchange reactions with chemi-ions such as H_3O^+ and $C_3H_3^+$. However, applying an appreciable electric field must destroy the space-charge neutrality of the combustion system because of electron depletion. Therefore, the recombination of chem-ions with electrons will be diminished, and the total positive-ion concentration will increase substantially. In turn, this necessarily will increase the number of large hydrocarbons participating in charge exchange reactions. Thus, ion–neutral nucleation, with a significantly larger rate constant, will contribute to a net increase in the number of nascent soot particles.

As the equivalence ratio decreases toward the soot formation limit, the relative number of chemi-ions in relation to large gas-phase hydrocarbons increases since the ion yield per fuel molecule also increases. The ionic–neutral nucleation route may well dominate under these conditions. However, a dramatic increase in ion concentration through electron depletion could elevate the chemi-ion concentration enough that the majority of soot precursors could be positively charged. For soot formation to occur, a large percentage of the nucleations must then be ionic–ionic in character with a resulting decrease in nucleation rate and total soot nuclei. A hypothetical formation mechanism from Wersborg et al. (*1*) is shown below. Here M and S represent large hydrocarbons and soot particles, respectively.

$$M + M \rightarrow S \tag{1}$$

$$M + M^+ \rightarrow S^+ \tag{2}$$

$$M^+ + M^+ \rightarrow S^{++} \tag{3}$$

$$S = S^+ + e^- \tag{4}$$

$$M = M^+ + e^- \tag{5}$$

$$M^+ + S \rightarrow M + S^+ \tag{6}$$

A detailed accounting of the reaction steps is not attempted here. However, several features of the mechanism are discernible. The formation route probably proceeds via the acetylene molecule as a building block to higher hydrocarbons. In methane combustion, it appears that the sequence is initiated by the methyl radical, CH_3, combining to give ethane, C_2H_6. Subsequently, hydrogen abstraction reactions finally yield acetylene, C_2H_2.

The other segment of the mechanism would include the formation of chem-ions, H_3O^+ and $C_3H_3^+$, which are the principal positive ions in hydrocarbon combustion. The formation of H_3O^+ is relatively well understood, and the reader is referred to the review article by Fontijn (*18*) for particulars. The same certainty cannot be applied to the reaction mechanism leading to $C_3H_3^+$. It appears, (*see* Kistiakowsky and Michael (*19*)), that the most probable reaction leading to this ion in the absence of oxygen is:

$$CH^* + C_2H_2 \rightarrow C_3H_3^+ + e^- \qquad (7)$$

However, all mechanisms leading to CH involve oxygen, e.g.:

$$C_2 + OH \rightarrow CO + CH^* \qquad (8)$$

to

$$C_2H + O_2 \rightarrow CH + CO_2 \qquad (9)$$

as reported by Gaydon (20) and Glass et al. (21), respectively. Bowser and Weinberg, however, observed substantial chemi-ion concentrations, even in the absence of oxygen. Either there is an alternate mechanism leading to CH, or CH is not the only chem-ion precursor. If C_2^* is the precursor, as suggested by Bowser and Weinberg through the reaction:

$$C_2^* + CH_3 \rightarrow C_3H_3 + e^- \qquad (10)$$

how is it formed?

One possible path would be through the reaction:

$$C_2H + H \rightarrow C_2 + H_2 \qquad (11)$$

as proposed by Jensen (22). It remains clear, however, that chemi-ionization is quite substantial, even in the absence of oxygen. This evidence lends itself to an explanation of the rather anomalous chemi-ion profile depicted in Figure 2.

Pyrolysis reactions are undoubtedly occurring after combustion, and chemi-ionization may be occurring in addition to charge exchange reactions between primary ions, such as $C_3H_3^+$, and light hydrocarbons. Hence, the chemi-ion profile would not decay necessarily as it does in leaner combustion, where recombination begins to dominate ion formation. Instead, a different sequence of chemi-ion reactions may become important, with the precursor being C_2, rather than CH.

The kinetics from here remain even more uncertain, but the poly-acetylenes clearly must acquire charge very early in the sequence since charge affects the soot formation from its inception. Other charging mechanisms are present almost certainly once soot nuclei are formed. The experimental ion traces must be evaluated cautiously because of the nonideal wave effects responsible for a pressure and temperature rise during the most rapid soot formation, i.e., the latter part of the fuel-lean combustion. The substantial increase in ionization approximately 1.5 msec after the shock passage cannot be explained by nonideal effects and kinetic arguments. Instead, once the soot is formed, thermionic emission commences almost certainly and subsequently augments the ion concentration.

Furthermore, in the time interval surrounding the rather complex nucleation from the gas-phase products to solid matter, processes such as thermionic emission contribute to the ion concentration in such a way that the recombination of chemi-ions is not noticeable in the gross positive charge structure. The mechanism discussed above is, of course, only speculative, and it is based on ion–molecule reactions suggested by others (1, 12). A more detailed picture of the particular ions involved must await mass spectrometric analysis.

Conclusions

The kinetic processes immediately proceeding soot formation in rich hydrocarbon combustion have been investigated using the reflected shock region of the conventional shock tube. Applied fields of moderate intensity perturbed the nucleation route in either of two ways. In very rich combustion, the applied field augmented the soot formation, whereas, under more lean conditions but still above the soot-forming limit, the field eliminated soot formation from its onset. The experimental results suggest that the postulated mechanism of Wersborg et al. (1) may be substantially correct and indicate that chemically formed ions, particularly under slightly rich conditions where they are present in significant quantity, participate in nucleation of soot.

Acknowledgments

The work was supported by an NDEA fellowship for T. W. Lester, and during the last phase of the experimentation, S. Wittig was assisted by the NATO Senior Scientist program. He gratefully acknowledges many helpful discussions with G. Wortberg, R. W. T. H. Aachen, on the subject.

Literature Cited

1. Wersborg, B. L., Yeung, A. C., Howard, J. B., Symp. (Int.) Combust. (Proc.) 15th (1975) 1439.
2. Wersborg, B. L., Howard, J. B., Williams, G. C., Symp. (Int.) Combust. (Proc.) 14th (1973) 920.
3. Hardesty, D. R., Weinberg, F. J., Symp. (Int.) Combust. (Proc.) 14th (1973) 907.
4. Lawton, J., Weinberg, F. J., "Electrical Aspects of Combustion," Clarendon Press, Oxford, 1969.
5. Howard, J. B., Symp. (Int.) Combust. (Proc.) 12th (1969) 877.
6. Lester, T. W., Zallen, D. M., Wittig, S. L. K., Combust. Sci. Technol. (1973) 1, 219.

7. Lester, T. W., Zallen, D. M., Wittig, S. L. K., *Recent Dev. Shock Tube Res. Proc. Int. Shock Tube Symp. 9th* (1973) 700.
8. Hand, C. W., Kistiakowsky, G. B., *J. Chem. Phys.* (1962) **37**, 1239.
9. Calcote, H. F., "Ionization in Hydrocarbon Flames," Fundamental Studies of Ions and Plasma *1, AGARD Conf. Proc.* (1965) **8.**
10. Lawton, J., Weinberg, F. J., *Proc. Roy. Soc. (London)* (1964) **A227**, 468.
11. Matsuda, S., Gutman, D., *J. Chem. Phys.* (1970) **53**, 3324.
12. Bowser, R. J., Weinberg, F. J., *Combust. Flame* (1976) **27**, 21.
13. Zallen, D. M., Hirleman, E. D., Wittig, S. L. K., *Symp. (Int.) Combust. (Proc.) 15th* (1975) 1013.
14. Schneider, K. P., Grönig, H., Z. *Naturforsch.* (1972) **12**, 1717.
15. Homann, K. H., *Combust. Flame* (1967) **11**, 265.
16. Wersborg, B. L., Howard, J. B., Williams, G. C., "Project Squid Tech. Report MIT-72-PU" (1972).
17. Wersborg, B. L., Fox, L. K., Howard, J. B., "Project Squid Tech. Report MIT-86-PU" (1974).
18. Fontijn, A., "Project Squid Tech. Report AC-13-PU."
19. Kistiakowsky, G. B., Michael, J. V., *J. Chem. Phys.* (1964) **40**, 1447.
20. Gaydon, A. G., "The Spectroscopy of Flames," 2nd ed., p. 166, Chapman and Hall, London, 1974.
21. Glass, G. P., Kistiakowsky, G. B., Michael, J. V., Niki, H., *J. Chem. Phys.* (1965) **42**, 608.
22. Jensen, D. E., "Combustion Institute European Symposium," p. 382, Academic Press, New York, 1973.

RECEIVED December 29, 1976.

12

Effect of Metal Additives on the Amount of Soot Emitted by Premixed Hydrocarbon Flames

A. FEUGIER

Institut Français du Pétrole, 1 & 4, Avenue de Bois-Préau,
Boite Postale n° 18, 92502 — Rueil-Malmaison, France

The addition of alkali and alkaline-earth metals leads to two simultaneous effects on the amount of soot emitted by premixed C_2H_4-O_2-N_2 flames: (1) a promoting effect caused by ions which accelerate the nucleation process of carbon formation and (2) an inhibiting effect caused by a chemical mechanism through which the additive produces hydroxyl radicals that remove soot. Depending on the kind of metal and the experimental conditions, one of these effects is greater than the other. The volume fraction of soot particles and the particle temperature were measured in a region where temperature and soot concentration are practically constant, without agglomeration process. Barium is the most efficient metal. The addition of transition metals was insignificant in this region, so their intervention occurs in the agglomeration phase by charge transfer processes.

Carbon formation in flames is an important problem which is still insufficiently understood, particularly in regard to the wide field of additives as was noted by W. J. Miller (*1*) in the last Symposium on Combustion. Some studies have been undertaken which suggest a chemimal mode of action for some metals (*2*). On the other hand, other authors have claimed that ions could also play a role in carbon formation (*3, 4*) and that they were responsible for the pro-soot and anti-soot effects of numerous metallic additives on the rate of soot formation in diffusion flames (*5*).

To gain more information about these two points, we have studied the influence of alkali and alkaline-earth metals on the amount of soot emitted by premixed hydrocarbon flames. The results indicate that in the carbon formation zone, where agglomeration is absent (in diffusion flames this process does exist), additives lead to two simultaneous effects on the amount of soot emitted: a promoting effect caused by ions and an inhibiting effect resulting from a chemical mechanism. Depending on the kind of metal and the experimental conditions, one of these effects is greater than the other.

However, under the same experimental conditions, the influence of some transition metals, especially manganese, turned out to be insignificant. This result suggests a mode of action for this kind of metal often used in certain industrial combustion devices.

Experimental

Method. Rich ethylene–oxygen–nitrogen flames are stabilized on a cylindrical burner which is enclosed in a vessel fitted with quartz windows. Metals are added to the flames by electrically heating a pearl of salt (nitrates or chlorides) introduced into the nitrogen flow, and their concentration is determined from loss of weight.

In the burnt gases, the volume fraction of soot particles in the absence (X_o) and in the presence (X) of additives as well as the particle temperature T_p (assumed to be equal to gas temperature (6)) are measured by means of an optical method. In this method the intensity $(I_\lambda)_o$ emitted by a cross-section of luminous gas and received by a monochromator is measured and compared with the intensity I_λ^o received through the same optical system by a calibrated tungsten ribbon lamp, with the filament image uniformly filling the monochromator entrance slit. Under the conditions we are interested in here, scattering is negligible, and absorption is relatively low. Assuming the particles to be spherical, it can be shown that the ratio $(I_\lambda)_o/I_\lambda^o$ under a logarithmic form is (7):

$$\ln \frac{(I_\lambda)_o}{I_\lambda^o} + \ln \lambda \epsilon' = \ln (6\pi A X_o l) - \frac{c_2}{\lambda} \left(\frac{1}{T_p} - \frac{1}{T^o} \right) \qquad (1)$$

where ϵ' = the emissivity of tungsten at the true temperature of filament T^o and at wavelength λ, A = a constant which depends on the refractive index of soot particles, l = flame thickness, and c_2 = 1.438 cm deg. The values of T_p and X_o are determined by plotting the left-hand side against $1/\lambda$ (7).

On the other hand, we found that the small amounts of additives added to the initial mixture do not modify particle temperature appreciably. So, if $(I_\lambda)_o$ and I_λ are the intensities emitted with and without additive present, respectively, we obtain $I_\lambda/(I_\lambda)_o = X/X_o$ provided that

**Table I.   Mixture Compositions, Experimental Gas Temperature T,
and Volume Fraction of Soot Particle X_0**

Mixture No.	Equivalence Ratio	N_2/O_2 Ratio	T $(°K)$	X_0
A	2.61	1.63	1840	1.2×10^{-7}
B	2.69	1.63	1795	3.8×10^{-7}
C	2.73	1.63	1780	7.0×10^{-7}
D	2.81	1.63	1760	1.0×10^{-6}
E	2.79	1.16	1830	2.2×10^{-7}
F	2.90	1.16	1800	5.0×10^{-7}
G	3.00	1.16	1750	1.6×10^{-6}
H	3.00	0.91	1785	1.1×10^{-6}
I	3.00	0.73	1845	6.1×10^{-7}

the wavelengths chosen do not interfere with the emission lines of metal
or some metallic species (mainly oxides and hydroxides).

The measurements were made in the primary zone of carbon formation, i.e., above the flame front, in a region where T_p and X_0 (or X)
are practically constant and where there is no agglomeration. This latter
point can be verified by Equation 1 which ceases to be valid when particle
size is larger than about 500 Å (7). The initial mixtures were chosen so
that X_0 could be varied in sufficient proportions. The mixture compositions used are listed in Table I.

Effect of Cesium, Potassium, and Sodium. With unseeded mixtures
having a relatively small soot concentration, curve $X/X_0 = f$ (total molar
fraction of metal M_0) continuously increases (Figure 1, mixture E). The
promoting effect of the three metals is thus revealed, and it is all the
greater as the ionization potential is low. But when X_0 increases, an
inhibiting effect can also be observed, even with cesium (Figure 1, mixtures B and D). The initial additions of this metal decrease the amount
of soot emitted, and curve $X/X_0 = f(M_0)$ falls to a minimum point whose
value is relatively small and then increases with a slope that becomes
more gradual as X_0 increases. The same phenomenon occurs with the
addition of potassium and sodium. For even higher values of X_0, there
is only a small inhibiting effect (Figure 1, mixture G).

Effect of Lithium, Barium, Strontium, and Calcium. The addition
of these metals always created a net inhibiting effect, except in some
cases with barium. In Figure 2 the curves continuously decrease with
strontium, calcium, and lithium while with barium, the ratio X/X_0 increases beyond a certain value of metal concentration. At this point the
promoting effect then becomes greater than the inhibiting effect, and the
value of X/X_0 may even be larger than one. However, for the higher
X/X_0 values, the inhibiting effect becomes relatively pronounced and
increases in the order lithium–calcium–strontium–barium.

Effect of Manganese. This metal is relatively effective in reducing smoke from gas turbine combustion systems (8) and furnaces (9). However, under the same experimental conditions as above, no significant effort was ever found with this metal or with iron and nickel. Furthermore, when oxygen was injected into the primary zone of carbon formation with a microprobe, the soot decrease profile did not change when manganese was added to the initial mixture. Hence this metal does not appear to stimulate the oxidation of small soot particles.

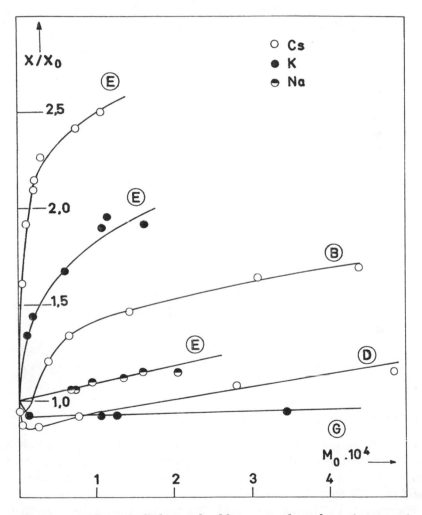

Figure 1.   Effect of alkali metal additives on the volume fraction of soot particles.   The circled letters refer to different mixtures that are identified in Table I.

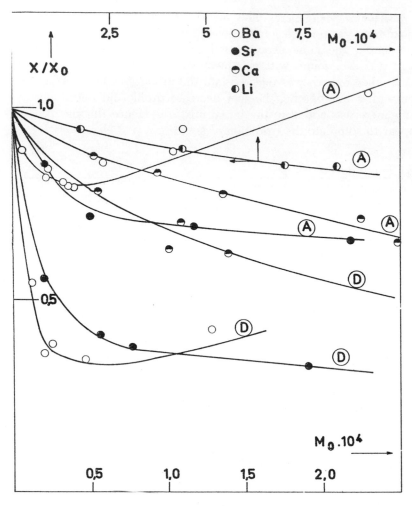

Figure 2.   Effect of alkaline earth metal additives on the volume fraction of soot particles. The circled letters refer to different mixtures that are identified in Table I.

Discussion

All the experimental results for alkali and alkaline-earth metals can be interpreted by considering that soot precursors P (e.g., polyacetylenes in the combustion of aliphatic hydrocarbons (*10*)) are both oxidized by OH radicals with the rate V_1 and are turned into small soot particles with the rate V_2. Depending on the kind of metal and the experimental conditions, one or the other of these reactions will be promoted, with the rates becoming V_1' and V_2', respectively.

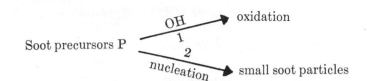

As a first approximation, in the flame zone of interest here, we can write:

$$X_o \propto P \frac{V_2}{V_1 + V_2} \quad \text{and} \quad X \propto P \frac{V_2'}{V_1' + V_2'} \quad \text{(2a and 2b)}$$

so that:

$$\frac{X}{X_o} \propto \frac{V_2'}{V_2} \frac{V_1 + V_2}{V_1' + V_2'} \quad \text{(3)}$$

Cesium, Potassium, and Sodium. With these three elements the inhibiting effect was relatively low, and the promoting effect was related to the ionization potential of the metal. From this, we will assume that $V_1' = V_1$. But to take into consideration the slight inhibiting effect, we will write $(X/X_o)_{\text{prom}} = (X/X_o)_{\text{exp}} + B$, with the value of B being slightly higher than the value corresponding to the minimum of curves $X/X_o = f(M_o)$.

At the same time, V_2' must depend on the concentration of thermal ions produced by the additive metal. We found that V_2' obeys the following equation:

$$V_2' = V_2 (1 + \alpha M^+) = V_2 (1 + \gamma \alpha M^+_{\text{eq}}) \quad \text{(4)}$$

where M^+ is the mean molar fraction of ions in carbon formation zone, and M^+_{eq} is the molar fraction of ions at thermodynamical equilibrium at the same temperature. Thus Equation 3 becomes:

$$y = \frac{X}{X_o} - 1 + B = \frac{(V_1/V_2)\gamma M^+_{\text{eq}}}{\gamma M^+_{\text{eq}} + (V_1 + V_2)/V_2 \alpha} \quad \text{(5)}$$

or

$$\frac{y}{(V_1/V_2) - y} \frac{1 + (V_1/V_2)}{(K/1 + \phi)^{1/2}} = \gamma \alpha (M_o)^{1/2} \quad \text{(6)}$$

where K is the equilibrium constant of the ionization reaction $M \rightleftarrows M^+ + e$, and $\phi = (OH)_{\text{eq}}/K'$ with K' the equilibrium constant of the reaction

$M + OH \rightleftarrows MOH$. The values of $(OH)_{eq}$, K, and K' were calculated from data in the literature (11, 12).

Figure 3, which corresponds to the results in Figure 1 (mixture E), shows that Equation 6 checks out fairly well. It implies that $V_1/V_2 = 2.30$ which, according to Equation 2a, is equal to $(P/X_o) - 1$ and, according to Equation 5, corresponds to the extreme value of the term $(X/X_o) - 1$ when $V_2' >> V_1'$ ($= V_1$). For each initial mixture, these extreme values could be attained experimentally if sufficient amounts of metal could be added, with everything else remaining the same. Figure 3 gives some other examples illustrating Equation 6. For all of the experi-

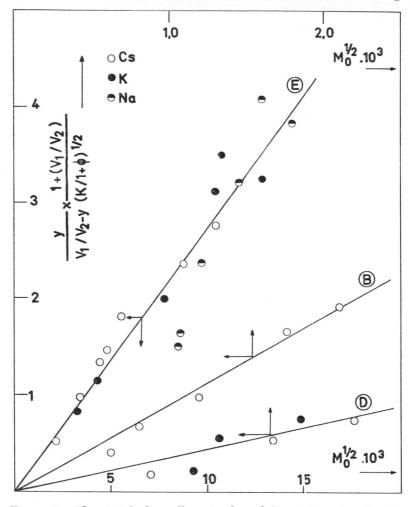

Figure 3.   The straight lines illustrate the validity of Equation 6, with $V_1/V_2 = 2.30$ (mixture E), 1.30 (mixture B), and 0.80 (mixture D). The circled letters refer to different mixtures that are identified in Table I.

mental conditions in which there is a promoting effect, V_1/V_2 varies between about 2.30 and 0.80, and $\gamma\alpha$ between 2.7×10^7 and 0.35×10^7.

Lithium and Alkaline–Earth Metals. With these elements X/X_o is usually less than one, and the inhibiting effect increases with X_o. For these particular values of X_o, we will assume that the inhibiting effect is the only one that exists, i.e., $V_2' = V_2$. Furthermore, as was suggested by Cotton et al. (2), we will assume that the additive metal reacts with some species of the flame to produce hydroxyl radicals which rapidly remove soot. It is well known that lithium easily forms the hydroxide LiOH via the reaction $Li + H_2O \rightarrow LiOH + H$. Likewise, alkaline-earth metals can produce hydrogen atoms via equivalent reactions; $M + H_2O \overset{1}{\rightarrow} MOH + H$, eventually followed by, $MOH + H_2O \overset{2}{\rightarrow} M(OH)_2 + H$. Then the hydrogen atoms can produce OH radicals by the reaction $H + H_2O \overset{3}{\rightarrow} OH + H_2$, which is known to occur at the end of the reaction zone in rich flames (10).

We found that V_1' obeys the following equation:

$$V_1' = V_1 + \mu M \tag{7}$$

where M is the mean molar fraction of free metal in the carbon formation zone. Thus Equation 3 becomes:

$$\frac{X_o}{X} - 1 = \frac{\mu M}{V_1 + V_2} = \frac{\mu \delta M_{eq}}{V_1 + V_2} \tag{8}$$

where M_{eq} is the molar fraction of free metal at thermodynamical equilibrium at the same temperature. This parameter is calculated by considering the different equilibria involving MOH, $M(OH)_2$, MCl_2 (alkaline-earth chlorides were used) and HCl, where the equilibrium constants are available in the literature (13, 14, 15).

Figures 4 and 5 illustrate the linear relation conforming to Equation 8. Such a relation is interpreted by applying the stationarity principle (with and without additive) for the radicals OH, H, and eventually MOH (if Equation 2 occurs), using Equations 1, 3 (eventually 2) and the unspecified reactions for the formation of hydrogen atoms and the disappearance of OH (mainly by CO and H_2). It can be easily deduced that:

$$(OH) = (OH)_o + \beta k_1 (M)$$

or
$$\tag{9}$$

$$(OH) = (OH)_o + 2\beta k_1 (M) \text{ if Equation 2 occurs}$$

where $(OH)_o$ and (OH) = the (OH) concentration without and with additive present in the initial mixture, and k_1 = the rate constant of

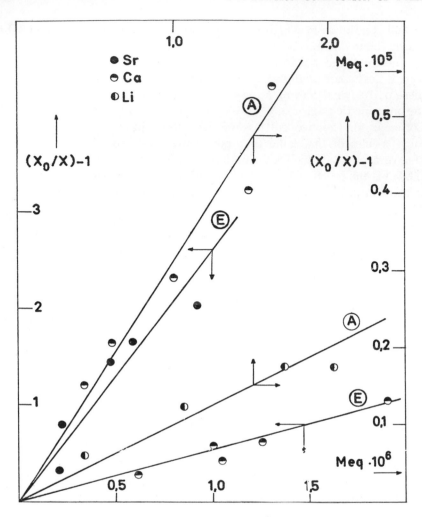

Figure 4.   The straight lines illustrate the validity of Equation 8. The slopes are proportional to the rate constant of the reaction: $M + H_2O \xrightarrow{1}$ *MOH + H. Circled letters refer to mixtures identified in Table I.*

Equation 1. Under these conditions, V_1' which is proportional to (OH) equals $V_1 + \mu M$ with μ proportional to k_1 (or $2\ k_1$). We found that Equation 2 does not have to be considered for the initial mixtures G, H, and I. (At thermodynamical equilibrium, moreover, the concentrations of $M(OH)_2$ are extremely low.) Equation 2 only comes into play for barium with the initial mixtures A, B, C, D, and E.

When only the inhibiting effect exists, a comparison of the slopes of the straight lines obtained with each metal gives for all of the experimental results:

$$k_{1(Ba)} \simeq 20\, k_{1(Sr)} \simeq 100\, k_{1(Ca)} \simeq 4 \times 10^3\, k_{1(Li)}$$

When inhibiting and promoting effects exist simultaneously, the general expression, Equation 3, with V_1' and V_2' given by Equations 4 and 7, respectively, leads to the following equation:

$$\alpha M^+ = \frac{(X_o/X)_{inh.} - (X_o/X)}{(X_o/X) - [1/(1 + V_1/V_2)]} = Q \qquad (10)$$

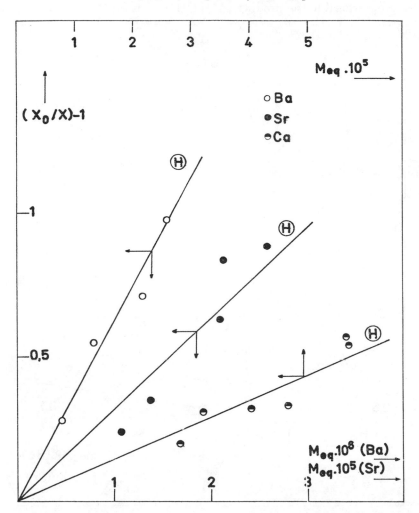

Figure 5.   The straight lines illustrate the validity of Equation 8. The slopes are proportional to the rate constant of the reaction: $M + H_2O \rightarrow MOH + H$. The circled letters refer to different mixtures that are identified in Table I.

where $(X_o/X)_{inh.}$ is the value of X_o/X if there is only an inhibiting effect. This value can be determined from the results obtained with calcium for the same mixtures and from the comparative values mentioned above.

As we have already seen, the promoting effect is caused by ions present in the carbon formation zone. The ionization potential of alkaline-earth metals is relatively high, so the only source of ions is in the chemi-ionization reaction $M + OH \overset{4}{\rightarrow} MOH^+ + e$. Indeed, we find that M^+ is proportional to the product $(M)(OH)$. This can be seen in Figure

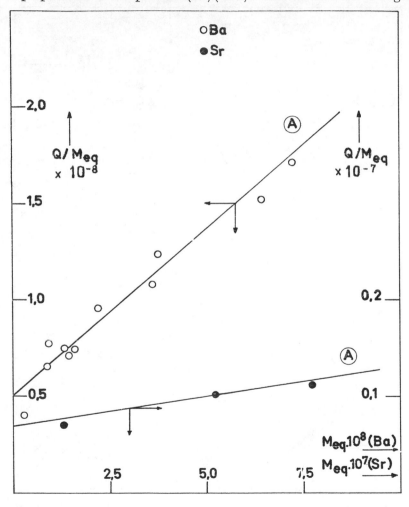

Figure 6.   The straight lines illustrate the validity of Equation 11. The ordinates in the beginning are proportional to the rate constant of the reaction: $M + OH \overset{4}{\rightarrow} MOH^+ + e$. The encircled letters refer to different mixtures that are identified in Table I.

6 in the linear relations obtained with barium and strontium according to the following equation:

$$\frac{Q}{M_{eq}} = \alpha\delta k_4 \,(OH) = \alpha\delta k_4 \,(OH)_o + \alpha\delta^2\beta k_4 k_1 \,(\text{or } 2\,k_1)M_{eq} \quad (11)$$

The ratio of the ordinates in the beginning for barium and strontium gives $k_{4(Ba)}/k_{4(Sr)} = 70$, which leads to: $E_{4(Sr)} - E_{4(Ba)} \simeq 16{,}000$ cal/mole (E = activation energy), in agreement with data published by Jensen (16).

Conclusions

Positive proof is given that thermal ions or chemi-ions accelerate the nucleation process of carbon formation. On the other hand, metals which easily produce hydroxides can inhibit carbon formation, and of these, barium is the most efficient. Transition metals such as manganese do not seem to play a role either in the formation phase of small soot particles or in their oxidation phase. We suggest that in industrial combustion devices, their intervention occurs in the agglomeration phase from involatile oxides formed in poor combustible zones. These oxides produce positively charged solid particles which can transfer their charges to the small soot particles, and consequently they prevent the agglomeration process.

Literature Cited

1. Miller, W. J., *Symp. (Int.) Combust. (Proc.) 14th* (1973) 307.
2. Cotton, D. H., Friswell, N. J., Jenkins, D. R., *Combust. Flame* (1971) **17**, 87.
3. Lawton, J., Weinberg, F. J., "Electrical Aspects of Combustion," Ch. 7, Oxford, 1969.
4. Ball, R. T., Howard, J. B., *Symp. (Int.) Combust. (Proc.) 13th* (1971) 353.
5. Bulewicz, E. M., Evans, D. G., Padley, P. J., *Symp. (Int.) Combust. (Proc.) 15th* (1974) 1461.
6. Kuhn, G., Tankin, R. S., *J. Quant. Spectrosc. Radiat. Transfer* (1968) **8**, 1281.
7. Feugier, A., *Combust. Flame* (1972) **19**, 249.
8. Giovanni, D. V., Pagni, P. J., Sawyer, R. F., Hughes, L., *Combust. Sci. Technol.* (1972) **6**, 107.
9. Salooja, K. C., *J. Inst. Fuel* (1972) **45**, 37.
10. Homann, K. H., *Combust. Flame* (1967) **11**, 265.
11. Ashton, A. F., Hayhurst, A. N., *Combust. Flame* (1973) **21**, 69.
12. Cotton, D. H., Jenkins, D. R., *Trans. Faraday Soc.* (1969) **65**, 1537.
13. *Ibid.* (1968) **64**, 2988.
14. Schofield, K., Sugden, T. M., *Trans. Faraday Soc.* (1971) **67**, 1054.
15. JANAF Thermodynamical Tables, The Dow Chemical Company, Midland, Mich., August 1965.
16. Jensen, D. E., *Combust. Flame* (1968) **12**, 261.

RECEIVED December 29, 1976.

13

Oxidation of Soot in Fuel-Rich Flames

F. M. PAGE and F. ATES

University of Aston in Birmingham, Gosta Green, Birmingham, England

The oxidation of soot in fuel-rich premixed flames proceeds through hydroxyl attack, with one atom of carbon removed by each OH radical hitting the surface of the soot. The absolute level and composition dependence of the disappearance rate of soot show that OH is the important species. The oxidation was followed experimentally by laser light scattering from soot particles generated in one flame and passed directly into the gas supply of a larger hydrogen/air flame controlled at 1700–2200 K.

The best known case of solid particles in a flame is ordinary soot, which is generated when a hydrocarbon is burned under fuel-rich conditions. Frequently such soot is consumed within the generating flame, and it is of interest to determine the mechanism of such consumption, be it a depolymerization (volatilization) or chemical reaction. Many workers have studied this phenomenon (1, 2), but the diversity of workers is rivalled only by the diversity of techniques and of theories, and a great bulk of the work refers to the slightly unreal conditions of massive (up to 2-cm diameter) spheres in oxidant-rich gases. The advent of the simple, high-intensity laser source improves considerably the light scattering techniques, which are particularly sensitive to particle size, and it seemed opportune to reexamine the consumption of soot in a closely controlled, fuel-rich environment using this approach.

Burner Complex

Kinetic studies in flames require a well defined environment, and particular attention was given to the design of the burner (Figure 1). Soot was generated in a primary burner following Fenimore and Jones (3), and the soot-laden gases were cooled and injected into the base of a secondary burner. Both primary and secondary burners were constructed from spaced hypodermic tubing, and the tubes for the observa-

0-8412-0383-0/78/33-166-190$05.00/0 © American Chemical Society

tion flame were surrounded by a further double row of tubes. In the primary burner, the center tubes were fed with metered supplies of ethylene, nitrogen, and oxygen, and the flame was ignited by a spark and was stabilized by a slow stream of air passing through the outer tubes. The conical chimney was maintained at 400 K by a heating tape to prevent condensation, and the cooled gases entered the bottom chamber of the secondary burner past a radial diffuser. The bulk of the gases for the secondary $H_2/N_2/O_2$ flame entered this chamber tangentially to assist in mixing and passed up the central tubes of the burner. A similar gas mixture was fed to the outer tubes so that the observation flame was protected from the atmosphere by a sheath of flame of the same composition and temperature. The whole double burner assembly could be raised or lowered to allow measurements to be made at different heights in the flame and therefore at different times. The temperatures of the flames were determined by the sodium D-line reversal method and were accurate to ±20 K. The gases from the primary burner were less than

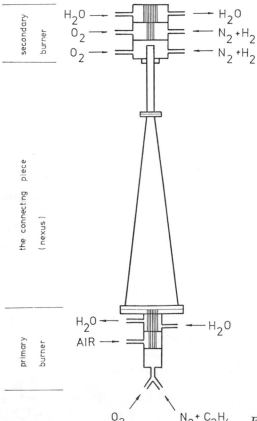

Figure 1.   Double burner system

10% of the gas flow to the central flame, and their composition was determined by gas chromatography or (for H_2O and solid carbon) by direct weighing.

Optical System

The light source used was a Ferranti 1-mW gas laser type SL3 radiating at 632.8 nm, and the scattered light intensity at an angle of 32° was measured by an EMI 9658B multiplier photocell through a Grubb–Parsons interference filter tuned to the laser light. The laser beam was chopped before entering the flame, and a reference photocell was activated by the same chopper. The output for the 9658B and reference photocells were passed to a Brookdeal 9412 phase-sensitive detector which drove a potentiometric recorder. This arrangement has been modified slightly to eliminate instabilities, and the reference signal is now taken from the unscattered laser beam. Even so, there remains a slow drift from the gradual deposition of soot within the burner system. The system was allowed to stabilize for 30 min to reduce the drift to acceptable limits. Care was taken to randomize the order of the heights at which observations were made and to check repeatedly at a standard height, usually 4 cm above the burner surface.

Analysis of Results

The intensity of the light scattered by a small particle is proportional to the sixth power of its radius, and the total scattered intensity will be obtained by summing over all particles. In order to simplify the analysis of the results, it has been assumed that all particles are consumed at similar rates, so that the number does not change, and that the particles are of a similar size, so that the summation may be omitted. The measured intensity I is expressed in terms of the particle number density (n) and the radius (a) as:

$$I = \alpha n a^6$$

Although soot is polydisperse (5), it probably approaches a self-preserving size distribution (6) so that the use of this simplified expression is valid.

The rate of change of intensity with height (h) may be related to the temporal rate through the burner gas velocity (u):

$$u \, d \ln I/dh = d \ln I/dt = 6 \, d \ln a/dt$$

If the attacking species X has a partial pressure [X] the number of effective collisions with the particle surface in unit time will be $kA[X]$, where A is the surface area of the particle, and if each effective collision removes a carbon atom, this will be the rate of removal of carbon. Alternatively, this rate may be expressed as a rate of change of particle volume through the mass of a carbon atom (m) and the density (ρ) which may be assumed to be that of graphite:

$$-dV/dt = km\,A\,[X]/\rho = -A\,da/dt$$

Expressing the volume change in terms of the radius and eliminating the total area of the particles leads to:

$$d\ln I/dh = -6\,km\,[X]/ua\rho$$

substituting $a = (I/an)^{1/6}$ gives:

$$I^{1/6}\,d\ln I/dh = \{6\,km\,(an)^{1/6}/up\}\,[X] = -\beta[X]$$

and replacing the slowly varying function $I^{1/6}$ by its mean value during a run $(\overline{I})$ gives the relation:

$$\beta\,[X] = -(\overline{I})^{1/6}\,d\ln I/dh$$

Results

The values of the function $\beta\,[X]$, measured for a set of four isothermal flames at 2060 K are shown in Table I, together with the value of β, assuming that X is in turn H, OH, O, and O_2 at the equilibrium concentration.

Table I.
The Rate Constant β, Assuming Various Attacking Species

$\beta[X]$	$\beta(H) \times 10^{-3}$	$\beta(OH) \times 10^{-5}$	$\beta(O) \times 10^{-7}$	$\beta(O_2) \times 10^{-8}$
24.2	17.5	2.4	7.6	10
15.4	8.0	5.8	26.1	35
14.8	7.3	5.8	27.4	217
59.3	47.1	3.3	7.9	6

Of the four possible reactant species, attack by the hydroxyl radical yielded the most nearly constant value of β, but the constancy is unimpressive. It is true that the use of equilibrium partial pressures will be in error, but the disequilibrium parameter (γ) of Bulewicz, James, and Sugden (4) depends more on temperature than on composition and should not affect the picture from an isothermal set of flames.

An alternative approach is to calculate the constant β from elementary kinetic theory, identifying the intensity at zero height in the flame

(2.5 in the arbitrary units used) with scattering from particles of the size seen in electron micrographs (2.5×10^{-9} m) of collected material.

If the molecular weight of the attacking species is M, the value of k at 2060 K is $6.6 \times 10^{27} M^{\frac{1}{2}}$ collisions sec^{-1} m^{-2}, and the density of graphite is 2250 kg m^{-3}, so that

$$\beta = 8.1 \times 10^5$$

Clearly this agrees most closely with the experimental value found assuming that hydroxyl was the effective species.

The value of β thus calculated would be decreased if not every collision were effective while the presence of radicals in excess of their equilibrium concentration would lower the experimental values. In the flame under discussion at 2060 K, the accommodation coefficient would have to be as low as 0.01 if hydrogen atoms were to be responsible, and the disequilibrium parameter would have to be as high as 17 if oxygen atoms were responsible and even larger for oxygen molecules. Since Bulewicz, James, and Sugden showed that the disequilibrium parameter γ is only about 1.3 at the point of measurement in these flames (4 cm), the latter possibilities may be ruled out. Such a low accommodation coefficient would be most unexpected on a rough and reactive surface, and only the hydroxyl radical is a likely attacking species.

Table II.

Flame	Percentage Composition			Temperature (K)
	H_2	O_2	N_2	
1	37	5	58	1690
2	30	8	62	1700
3	36	9	55	1860
4	32	11	57	1900
5	61	2	37	1900
6	40	8	52	1970
7	43	7	50	1970
8	46	6	48	1970
9	43	10	47	1980
10	60	5	35	2020
11	50	8	42	2030
12	37	12	51	2040
13	65	7	28	2050
14	50	10	40	2060
15	70	4	26	2060
16	76	4	20	2060
17	43	12	45	2080
18	60	8	32	2100
19	60	10	30	2120
20	50	12	38	2170

Effect of Disequilibrium

Cool hydrogen flames have an excess of radicals (H atoms, OH radicals etc.) over the calculated equilibrium values, and fairly small amounts of hydrocarbon fuel reduce these radical excesses. These excesses, expressed as the disequilibrium parameter ($\gamma = [H]/[H]_{eq}$) vary between flames and must be taken into account in making comparisons over a series of flames. An examination of the original paper (4) indicates that over a large number of hydrogen flames, the value of γ is determined only by the temperature and not by the composition, and it is possible therefore to use this data to compare results in the present work. The experimental determination of γ in sooty flames is difficult, and at this stage it is preferable to use values determined in clean flames while remembering that they are upper limits. Table II presents the values of $\beta[X]$ determined during the present study in 20 flames, together with the appropriate flame parameters, and the value of β calculated on the assumption of equilibrium ($\beta[X]/[OH]_e$) and disequilibrium ($\beta[X]/\gamma[OH]_e$) in the radical concentration. Attack by hydroxyl radicals at their equilibrium concentration is inadequate to explain the consumption of soot over a wide temperature range, but all observations can be explained if the hydroxyl radicals are, as expected, in excess of their equilibrium level.

Values of β [X]

γ	$\beta[X]$	$\beta[X]/[OH]_e \times 10^{-5}$	$\beta[X]/\gamma[OH]_e \times 10^{-5}$
16.6	8.6	61.8	3.7
15.5	11.5	33.8	2.2
4.2	18.5	9.7	2.3
3.2	26.2	5.3	1.6
3.2	2.9	8.2	2.5
2.1	17.2	4.3	2.0
2.1	15.1	4.7	2.2
2.1	14.6	5.6	2.7
2.0	22.3	3.8	1.9
1.6	13.8	5.3	3.3
1.5	16.6	2.9	1.9
1.4	45.0	2.8	2.0
1.33	15.7	3.1	2.4
1.25	24.2	2.4	1.9
1.25	15.4	5.8	4.6
1.25	14.8	5.8	4.6
1.25	59.3	3.3	2.6
1.0	19.0	2.1	2.1
1.0	23.4	1.7	1.7
1.0	58.6	1.9	1.9

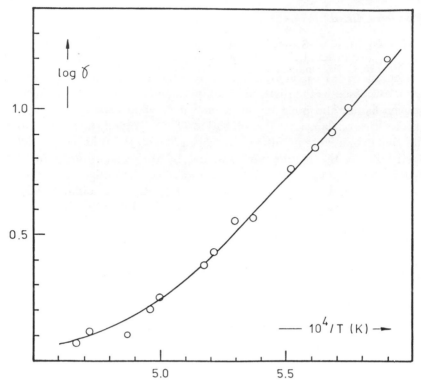

Figure 2.   Estimation of the disequilibrium parameter (γ)

Conclusion

The consumption of soot in a low temperature, fuel-rich, laminar, premixed flame can be accounted for quantitatively if attack is by hydroxyl radicals present in the expected excess over their equilibrium concentration. Only one in four collisions of the hydroxyl radical with the surface need be effective to account for the observed rate. Other attacking species have been considered, but the pattern of rate of attack does not agree with hydrogen's being important, and the very low concentration of oxygen, either in atomic or molecular form, makes it impossible to account for the observed rates, even allowing for radical excess. This is not to say that in fuel-lean flames, where the concentration of these species is much greater and where hydroxyl is not greatly altered, that the dominant mode of attack may not be caused by molecular oxygen, but the efficiency of attack by hydroxyl is so high that such dominance will be caused solely by the concentration of species.

Literature Cited

1. Mulcahy, M. F. R., Smith, I. W., *Rev. Pure Appl. Chem.* (1969) **19,** 81.
2. Park, C., Appleton, J. P., *Combust. Flame* (1973) **20,** 369.
3. Fenimore, C. P., Jones, G. W., *J. Phys. Chem.* (1967) **71,** 593.
4. Bulewicz, E. M., James, C. G., Tugden, T. M., *Proc. Roy. Soc.* (1956) **A235,** 89.

RECEIVED December 29, 1976.

14

Noninterfering Optical Single-Particle Counter Studies of Automobile Smoke Emissions

S. L. K. WITTIG[1] and E. D. HIRLEMAN

School of Mechanical Engineering, Purdue University,
West Lafayette, IN 47907

J. V. CHRISTIANSEN

Department of Fluid Mechanics, The Technical University of Denmark,
2800 Lyngby, Denmark

A ratio-type counter with associated electronic data acquisition equipment has been designed, built, and calibrated for in situ, real time measurements of automobile exhaust gas emissions. The system design was based on theoretical considerations of light scattering and associated error analysis. Preliminary calibration measurements were conducted using a Collison-type atomizer to generate particle clouds of different size distributions. For a first application a four-cylinder engine using both leaded gasoline fuel as well as a 15% methanol blend was used. The results deviated considerably from previous data obtained with conventional sampling probe and impactor systems. The influence of methanol blends and of acceleration effects is discussed and qualified.

In the past considerable attention has been directed towards the well publicized formation, analysis, and control of gaseous air pollutant emissions such as NO_x, CO, unburned HC, and SO_x. In contrast relatively little effort has been devoted to the study of nongaseous emissions, especially particulate emissions from internal combustion engines. This is partly caused by the fact that, on a weight basis, the amount of relatively large particulate matter formed in and emitted from automobiles

---

[1] New address: Institut für Thermische Strömungsmaschinen, Universität Karlsruhe (TH), Postfach 6380, D-7500 Karlsruhe 1, West Germany.

and other sources is somewhat smaller than that of the gaseous components.

In the last few years, however, it has been recognized that particulates have a far greater impact than previously assumed. Particulates include a variety of substances which may be toxic in their own right, they exhibit detrimental effects on machine elements, and they participate in the chemical processes of the combustor as well as of the atmosphere. The obvious need to minimize particulate emissions requires the detailed understanding both of the particle formation and emission processes in continuous combustors, such as gas turbines as well as reciprocating engines, and of techniques for monitoring the ability of individual engines to meet particulate emission standards. Neither of these are presently available.

Very recently smoke emissions from aircraft gas turbine combustors and particularly from diesel engines have been investigated. Restrictions are now being placed on exhaust smoke opacity, but the effects of the smoke, other than plume visibility, are still not well understood (1). Instruments which are currently used to monitor visible smoke include the Bosch Spot, Hartridge, and USPHS smokemeters. In the Bosch meter, the reflectance of a known volume of exhaust gas passing through a filter is measured. In the other two techniques, the attenuation of a light beam as it passes through the exhaust is determined. In all three cases, however, the response of the instrument depends upon an integrated function of the exhaust particle size distribution, particle refractive index, and spectral distribution of the light source (1). However, such relatively crude techniques do not produce the desired detailed insight into formation mechanisms, particulate size distributions, coagulation history, and other aspects.

The formation of soot and smoke particulates during combustion in the internal combustion engine is not well understood. Heterogeneous as well as homogeneous processes such as pyrolysis, polymerization of vapor phase fuel molecules, etc. may play important roles. Nucleation, condensation, and coagulation into the final emitted particles are the major steps (2). The smoke particles are acted upon by many physical and chemical processes, most of which are important in the cylinder as well as in the exhaust pipe. The aerosol is thus a highly dynamic system, and considerable care must be taken to prevent falsification during sampling.

These considerations are particularly important in the area of alternative automotive fuel combustion. Methanol/gasoline mixtures are now recognized as an important alternate fuel candidate for use in conventional engines as well as for future automobile power plants (3). Large scale experiments are being conducted in Germany and the U.S., and detailed studies have been made at the Laboratory for Energetics, DTH (4). Based on preliminary results, it can be assumed that methanol tends to reduce various pollutants, but it does not reduce aldehydes. One aspect, however, which has not been considered so far is the emission of particulates, especially in the low- and submicron range. Therefore, one aspect of the present study was to investigate the details of the particulate emission behavior of engines using alternate fuels.

As indicated before, previous detailed studies of particulate formation and emission processes in automobile engines and exhaust systems have been done using probes which extracted a (hopefully) representative sample of the exhaust gas that was subsequently analyzed for various particulate properties (5). Similar techniques have been used in the most recently developed systems to measure the amounts of particles emitted from auto exhaust. Such sampling techniques, however, suffer from inherent disadvantages in that they interfere with the hot flow particulate processes and have an unknown effect on the particulates as they flow through the system to an exterior point of analysis. This problem can also be compounded by several phenomena: anisokinetic sampling, deposition of particulates in the sampling system, or condensation caused by temperature drops.

An alternative is the use of an optical method to measure particulate concentrations and size distributions. This technique has the obvious advantage of having a negligible effect on the particulates since the equipment would be external to the exhaust system. An optical method also has the potential to be much simpler to use since it would eliminate the need for elaborate and cumbersome systems containing probes, stack samplers, flow development tunnels, filters, and heat exchangers. In addition, final data from an optical system could be immediately obtained electronically as opposed to weighing the various filters in a particle impactor by hand, and as such, the optical analyzer is a real time instrument capable of following exhaust gas fluctuations and other nonsteady effects.

Since particles in the micron and submicron size range are of primary importance, the emphasis in this study is on using the single-particle light scattering characteristics, and a new optical single-particle ratio counter was improved and calibrated. Furthermore, an electronic data acquisition system was developed, and finally the counter was applied to an in situ, real time study of automobile exhaust gas analysis.

Optical Particulate Measurements

Multiple Ratio-Type Single-Particle Counters (MRSPC). Diagnostic techniques using the light scattered by an individual particle passing through a highly illuminated region are based on the Mie theory. Here, the radiant flux (F) scattered at an angle θ from the incident light is:

$$F = F(I,\lambda,\theta,\alpha, n) \tag{1}$$

with I the incident light intensity, α the dimensionless particle circumference $(\pi d/\lambda)$, λ the wavelength of the incident light, and n the complex refractive index. The analysis requires the solution of the Mie intensity functions i_1 and i_2 corresponding to the two directions of polarization as discussed in Ref. 6. In the present study, a computer code to generate the solutions was made available by Dave (7).

As Equation 1 indicates, the flux scattered in a certain direction from a particle depends on both the particle properties and the intensity of incident light. Problems arise because of the Gaussian intensity distribution in a TEM$_x$ laser beam and fluctuations present in most light sources. A small particle, for example, passing through the center of the laser beam may conceivably scatter the same flux as a larger particle moving through an off-center point of lower intensity. Two methods for eliminating this problem have been suggested. The first is to aerodynamically focus the particle laden flow through a small portion of the laser beam where the intensity is approximately constant. Such a technique was used by Gebhardt, et al. (8) but obviously precludes in situ analysis. Hodkinson (9) originally suggested and Gravatt (10) investigated the concept of taking the ratio of light intensities scattered at two angles, effectively canceling out the incident intensity term in Equation 1.

A characteristic curve of intensity ratio as a function of particle diameter for measurement angles $\theta = 12°$ over $\theta = 6°$ and $n = 1.54 - 0i$ as calculated from Mie theory is shown in Figure 1. This particular index of refraction corresponds to a leaded petroleum-base aerosol particle (11). The effects of the complex component are demonstrated in a comparison

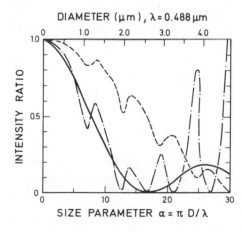

Figure 1.   Calculated light scattering ratios. (———) $n = 1.57 - 0.56i$ $(12°/6°)$; (—·—) $n = 1.54 - 0i\,(12°/6°)$; (– – –) $n = 1.54 - 0i\,(6°/3°)$.

for $n = 1.57 - 0.56i$, the index of refraction for flame-produced absorbing soot particles (12). The curve for nonabsorbing particles has numerous fluctuations whereas the absorbing particles have a smoother characteristic, but both have the same general shape.

Several problems are inherent in determining the particle diameter from a measurement of an intensity ratio. First, the calibration curve depends on the index of refraction, and therefore uncertainties are introduced when analyzing particles of unknown composition. Secondly, there is not a unique solution in general since several possible diameters correspond to any measured intensity ratio. These problems have been recognized (13, 14, 15), and we have added a detailed analysis (16).

When characterizing particulate matter of unknown composition, it is necessary to assume a value for the index of refraction to infer the diameter from a measured intensity ratio. This causes inherent uncertainties in any reported size distribution unless all particles are of a known and uniform composition. In the case of automobile exhaust particles, the composition is certainly unknown and would probably include some combination of carbon particles and lead halides condensed on nuclei. Figure 1 is indicative of expected variations from such a spread of particle compositions. In the range of low α, the characteristic curve for nonabsorbing particles oscillates around an average value which is approximately the $n = 1.57 - 0.56i$ data. Thus the intensity ratio curve for the absorbing soot is a convenient one to assume as the calibration standard for automobile exhaust particulates. Here this assumption results in a maximum error of approximately 30% when measuring particles of unknown composition.

Ideally a single-particle counter should have a response function monotonically dependent on particle size exemplified by that portion of Figure 1 for $n = 1.57 - 0.56i$ and $\alpha \leq 17$ but with zero response for all other α. The real situation however is quite different as indicated. Two types of deviations from desired behavior can be identified.

The first problem area occurs at $\alpha \simeq 7$ for $12°/6°$ ratio for weakly absorbing particles where, as exemplified by Figure 1, an "S" in the characteristic curve precludes a unique determination of α for a measured intensity ratio of 0.5. This results in an uncertainty of about 20% in determined diameter, roughly the same as that from index-of-refraction effects discussed above. In fact, the uniqueness problem adds no new uncertainties since the "S" phenomenon is covered by the uncertainty band introduced when the absorbing particle intensity ratio curve is assumed to be caused by unknown particle composition. Conversely, this problem must be dealt with when analyzing nonabsorbing particles of a known composition, for instance in a study of cooling tower droplet size distributions.

A second and more critical problem is caused by the nonzero response function for particles of α greater than the upper limit of the monotonic portion of the response curve ($\alpha \simeq 17$ for $12°/6°$). Sizes of particles greater than the applicable range of a RSPC would be classified incorrectly as that diameter within the valid range corresponding to the same scattering ratio. For example, a 3.8-μm particle with $n = 1.54 - 0i$ would produce a $12°/6°$ scattering ratio of about 0.75 and would therefore be sized as 0.7 μm diameter. Previous technical papers on RSPCs have completely neglected this problem, probably since it is concealed during calibration runs with monodisperse spheres. It is apparent, however, that conventional ratio counters would incur significant errors during analysis of the highly polydisperse aerosols emitted from combustion processes.

This dilemma could be resolved if the counter could identify particles outside the valid range and subsequently eliminate them from consideration. Such discrimination can be accomplished using a concept originating in the early stages of the present study (*17*): the light scattered at a third angle is monitored and provides a consistency check to insure that a particle is not larger than the limit of applicability for a particular counter design.

The previous discussion is quantitatively valid only for the angles and indices of refraction presented, but the analysis and conclusions would qualitatively hold for a general combination of angles and indices of refraction. Also, in practice the uncertainty introduced by the refractive index effect requires that a range of ratio combinations be used in the three-angle discrimination test rather than two exact ratios, but the advantages are still present. The method can also determine refractive indices when inversed.

An underlying assumption in the discussion up to this point has been that the particles are perfect spheres, which is probably not the case for the combustion-generated particulates which have passed the nucleation stage and have undergone coagulation processes. Any optical aerosol analyzer calibrated with spherical particles measures an optical equivalent diameter. In other words, it measures the diameter of a sphere of the calibration index of refraction that has the same light scattering characteristics as the particle being analyzed. This dimension would be important for radiation and visibility considerations, but the aerodynamic equivalent diameter controls some aerosol processes including lung deposition. Fortunately, optical measurements in the forward lobe are nearly independent of the particle shape and surface details and represent a relatively accurate measure of the mean projected area of the particle (*8*). This occurs because the forward lobe scattering approximates Fraunhofer diffraction whereas outside this domain, the dominant scattering mechanisms are refraction and reflection which depend more strongly on detailed particle structure.

Experimental Arrangement. The MRSPC optical particle counter used (*see* Figure 2) a 15-mW Argon-ion laser. The light beam (0.488 μm) emerging from the laser was about 1 mm in diameter with a divergence of less than 1.0 mrad. The beam was passed through a polarization rotator and was focused subsequently to a 66-μm diameter waist. Figure 3 is a photograph of the actual optical elements.

The receiving optics separated the light scattered from the sensitive volume at the angles 1°, 3°, 6°, 12°, and 24° for independent measurements. The system designed to accomplish this task is also shown in

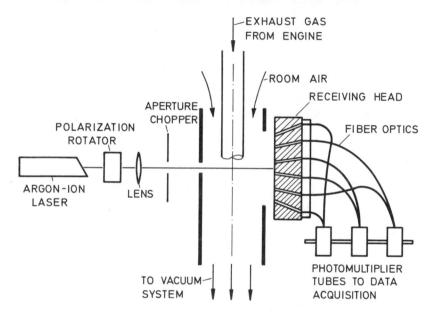

Figure 2.   Multiple ratio single-particle counter and exhaust system

Figure 3, with six conical sections. These were fit together separated by small wires, thereby allowing five conical sheets of light ($\Delta\theta \simeq 0.8°$) originating in the sensitive volume to pass through the cones. A receiving head with five annular rings of optical fibers corresponding to openings at the rear of the cone collected the scattered light. The optical fibers (Jena Glaswerk Co.) from each ring were then bunched together separately and encased in a metal plug so that the light passing through the fibers could be conveniently directed onto the photomultiplier tubes (EMI 9524B) by specially designed adaptors.

A data acquisition system converted output from the photomultiplier tubes into a format interpretable as particle size distributions.

Figure 3.   Optical components

Initially the signals were displayed on a fast storage oscilloscope, and a typical particle scattering trace for 3°, 6°, and 12° is shown in Figure 4. A nearly Gaussian intensity distribution was always observed, and the particle velocity can be estimated from the known beam width divided by the measured transit time.

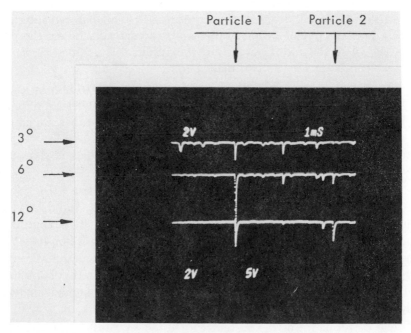

Figure 4.   Angular scattering signals

Numerous pulses are present on the 3° trace, but most of these either pass outside the smaller 12° field of view or do not scatter enough light at 12° to be detected. The gain of the 3° detector is significantly lower than for 6° and 12°; in fact a particle less than about 1 μm diameter would not be detected above the noise at 3° as the particle designated 2 on Figure 4. Conversely, a very large particle has a much narrower Mie forward lobe, and therefore could have a detectable 3° scattering signal even when the 6° and 12° signals are roughly of the same magnitude as for particle 1 of Figure 4. It is therefore possible to eliminate particle 1 from consideration since it is outside the valid range of the counter. However, this can be done only when the additional information supplied by the 3° data is available. In the absence of the third angle information, particle 1 would have been sized erroneously as within the first portion of the calibration curve of Figure 1.

The algorithm just discussed is a qualitative version of what must be done for each particle detectable above the noise level at both 6° and 12°. In reality the allowable 3° signal range depends on the observed 12°/6° ratio. Thus any data acquisition technique must be able to observe the 12°/6° ratio, to calculate a corresponding allowable 3°/6° ratio for comparison with that actually measured, and finally to decide on the validity of the particle based on this comparison.

Electronic Data Acquisition System. Essential for the effective use of a MRSPC is an electronic data acquisition system that can interface directly with the photomultipler tubes and automatically detect the particles, calculate the ratios, and store the results. The function of the interface can be divided in two main parts:

(1) A logic part responsible for selection of signals to be processed, control of analog switches, and interfacing time delay units

(2) An analog part where the signals from the PM-tubes are amplified, integrated, and ratioed.

A functional diagram of the system for one pair of angles is shown in Figure 5.

The details of the logic system are described elsewhere (*17*). However, in the start situation, the integrators have been reset, and the analog switches are in the positions shown on the diagram. By accepting a signal, the switches S1 and S2 connect the integrators with non-inverting preamplifiers, and the signals from the PM-tubes are integrated, the two integrated values are fed into the ratio circuit, and the ratio of the integrated signal multiplied by 10 is fed onto the multi-channel analyzer (MCA) by closing the switch S3 (compare also Ref. *10*).

The performance of the system was tested by substituting the PM-tubes with a pulse generator. These measurements gave an overall inac-

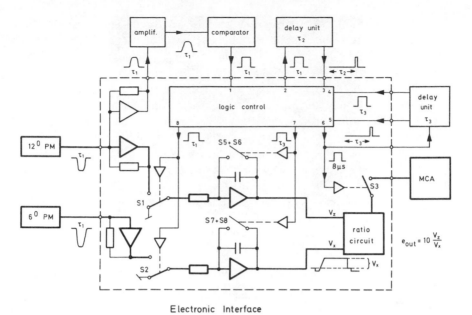

Electronic Interface

Figure 5.   Electronic data acquisition

curacy of 1–2% except for very small values of the integrated signal
where the inaccuracy increased to 10%.

Calibration. The task of any particulate analyzer is to measure the
particle number concentration as a function of diameter, or equivalent
diameter in the case of nonspherical particles. Results are generally
grouped into number densities for several discrete diameter ranges.
Ideally, the analyzing system should give a transducer output monotonically related to particle size, and a MCA can conveniently classify the
signals by level. The number of counts recorded in a particular channel
then represents the number of particles analyzed in the diameter range
corresponding to the voltage limits of that channel. A typical MCA output is shown in Figure 6, with counts per unit time on the ordinate and
voltage level on the abscissa. One purpose of calibration is to transform
the voltage levels to corresponding particle diameters so that Figure 6
is a measure of the normalized particle size distribution $f(d)$ with units
particles per unit time with diameters between d and $d + \Delta d$. The final
desired result is the particle concentration distribution $C(d)$ given by:

$$C(d) = N \frac{f(d)}{V} \tag{2}$$

where N is the total count rate of particles analyzed, and V is the volumetric sampling rate. The conversion in Equation 2 is not straightforward

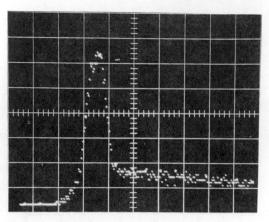

*Figure 6.   Multi-channel analyzer display for
glycerin particle calibration*

since V is generally not constant but depends on both the diameter and index of refraction of the particle being analyzed $[V = V(d,n)]$ as discussed by Hirleman et al. (*17*).

Consequently, raw output from a particulate analyzer in the form of Figure 6, even when the abscissa has been calibrated, is not necessarily an accurate representation of the particulate concentration distributions.

From Equation 2 it is apparent that there are two calibrations necessary for any particulate analyzer. First, the instrument must be calibrated to correctly determine the diameter of any particle "seen" by the analyzer in order to find $f(d)$. Secondly, the dependence of the volumetric sampling rate on particle diameter must be known in order

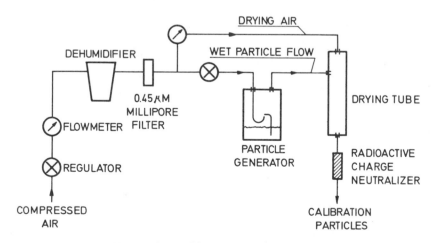

Figure 7.   Calibration droplet generator

to allow accurate calculation of $C(d)$. A detailed discussion of the techniques used for these two calibrations is presented in Ref. *14*. Principally, the system parameters for the MRSPC were determined by calibration using nearly monodisperse polystyrene latex spheres. As shown in Figure 7 the original latex slurries (10% polystyrene spheres in water) were further diluted about 1000 times using doubly distilled water filtered through a millipore 0.22-μm filter. These final solutions were atomized using a collison-type spray atomizer, and the water was subsequently evaporated off the latex spheres. The latex particles were then passed through a radioactive charge neutralizer (AM-241) to minimize coagulation and deposition of particles enroute to the counter. Electron microscope pictures of the 1.011-μm polystyrene spheres collected on a 0.45-μm millipore filter placed after the drying tube of the delivery system were also taken for control. A dilution ratio of 1000 was adequate to give single particles for the sphere sizes used in this study. Smaller particles, however, require higher dilutions. The results of a calibration with polystyrene spheres are shown in Figure 8 and are compared with

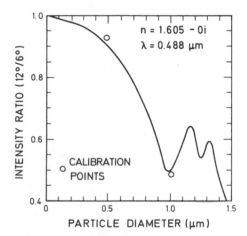

Figure 8.   Comparison of calibration points with theoretical predictions (polystyrene spheres)

the Mie theory. The ratios indicated are those for the measured mean values $\Delta d/d$ for calibration runs with monodisperse spheres averaged around 20%.

As previously indicated, a final determination of the particle size distribution requires knowledge of the sensitive volume which is actually a function of particle diameter and composition. The first step in quantifying the sensitive volume was to investigate the light intensity distribution near the laser focus. This was accomplished by traversing a 2.5-μm diameter wire through the beam using a micrometer screw accurate to 1 μm. The radial intensity distribution was found to be Gaussian with a beam waist diameter in the focal plane of 66 μm. Figure 9 shows the locations of the $1/e^2$ intensity points near the laser focus for a plane containing the z-axis. Also superimposed on Figure 9 are the fields of view of the collecting optics. The solid of revolution with a cross-section represented by the shaded area is the sensitive volume for points of incident intensity greater than I_{max}/e^2 and in the field of view of both 6° and 12°. This region was actually mapped out using the wire and found to have a volume of approximately 3×10^{-6} cm^3 with a frontal area to the flow of 5.2×10^{-3} cm^2.

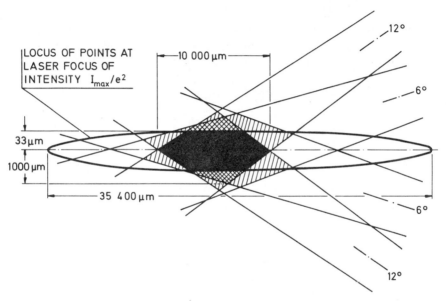

Figure 9.   Sensitive volume and optical field of view near laser focus

Figure 9 is not drawn to scale as the radial dimension is proportionately much smaller than the depth of focus along the beam axis. Thus many particles can pass through regions of lower intensity but still be in the field of view of both 12° and 6° detectors (cross-hatched region of Figure 9). Such considerations point out that the actual sensitive volume depends on the absolute scattering intensity of a particle since an extremely small particle might scatter with sufficient signal-to-noise ratio only at the very center of the shaded region whereas a larger particle could exceed the electronic threshold even when passing through the hatched region of lower intensity.

As previously indicated, determination of an actual particle size distribution requires an expression for the volumetric sampling rate V (or frontal area A_s × the flow velocity) as a function of particle diameter. This can be achieved if the instrument parameter $f \cdot l$ (f-number times the length of the sensitive volume) can be determined. Since f and l are the most uncertain because of spherical aberration of the lens and resolution deficiencies of the collecting optics, $f \cdot l$ was calibrated out by determining the actual frontal area at the $1/e^2$ intensity points as described above (Figure 9). For the counter under consideration the following relation holds:

$$A_s(d,n) = \frac{0.236}{\pi} \lambda \sqrt{\ln\left[\frac{i(d,n)}{22.4}\right]} \text{ mm}^2 \qquad (3)$$

From Equations 2 and 3 it can be seen that wider particle size distributions need to be corrected as shown. In the present study, an additional sensitive volume check was made using an electron microscope count of the particles on a millipore filter placed in the delivery tube for a fixed time. Results indicated that the analysis and observations were consistent. This also applies to calibration measurements with silicon oil mists.

Automobile Smoke Emission Measurement

The engine used was a standard four-cylinder Ford Cortina engine. It was mounted on a dynamometer to allow precise control of engine rpm and loading. The optical measurements were made behind 5.78-m length of 3.5-cm i.d. exhaust pipe which extended straight out from the side of a modified manifold. A high-capacity vacuum system was connected to a 10-cm i.d. cylindrical tube placed over the end of the exhaust pipe drawing the exhaust gas and a surrounding cylindrical sheet of air through the optical volume and into the vacuum system. The sensitive volume of the particle counter was situated 1.5 cm downstream from and centered in the end of the exhaust pipe. A schematic of the apparatus is shown in Figure 2. Chromel–Alumel thermocouples were used to measure temperature profiles along the exhaust pipe. The temperature at the observation point was varied using heating tape and insulation or forced convection cooling supplied by fans. A typical European blend regular fuel was selected for the experiments. It was obtained from Exxon (Esso) and contained 0.54 g/L lead.

Extensive studies of the size distribution of particulates emitted during a simulated highway cruise at 55 mph were completed. The exhaust and optical system configurations were maintained, but the heat transfer from the exhaust pipe was varied allowing detailed analysis of two exhaust gas temperatures—220° and 315°C—for constant exhaust and optical system configurations. The RSPC calibration was checked after the runs were completed. Figure 10 shows the observed particle distribution function vs. particle diameter, such that the area under the curve between any two diameters represents the number of particles in that particular diameter range per liter of gasoline consumed by the engine. The data were found using Equation 2 and the observed fuel consumption rate. Also pictured in Figure 10 are data taken by Ganley and

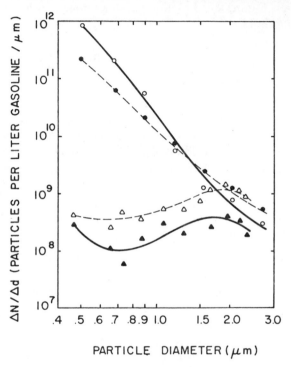

Figure 10. Particle size distribution for simulated 55-mph cruise. Ref. 5: (○) 208°C; (●) 298°C. Present study: (△) 220°C; (▲) 315°C.

Springer (5) using a sampling probe–impactor system for a similar cruise condition with an eight-cylinder Chevrolet engine. The reported impactor data, taken on a mass basis, was converted to a number basis here assuming a particulate density of 1.8 g/cm³ and normalized to gasoline consumption to facilitate comparison.

Large discrepancies are apparent between the two studies, and the only common trend is the increase of particle concentrations with decreasing temperature at the observation plane. The optical measurements indicate about three orders of magnitude fewer particles of small diameters than observed by Ganley and Springer (5); however, the concentrations of larger particles are nearly equal. Discussion of the differences in observed particle concentrations and size distributions can be separated into two major disciplines. First, the particle characteristics may actually have been different, and secondly, some deviations were undoubtedly introduced by the two measurement techniques.

Several factors may have caused an actual difference in particle characteristics between the two apparatus, even though Figure 10 was normalized to a common parameter, volumetric fuel consumption. Ganley and Springer (5) took detailed data using Indolene HO 30 research fuel containing 0.87 g/L lead, compared with 0.54 g/L lead in the fuel used here. Comparative tests suggested that Indolene HO 30 would probably produce more particles in the diameter range 0.3–1 μm than regular fuel at high temperatures (4). Thus the combined effects of fuel composition and added lead could account for some of the variations.

Another major factor is the difference in engine parameters and exhaust system configuration. Exhaust pipe diameter and length from the engine to observation plane effect the important mechanisms of coagulation, deposition, and re-entrainment of particles thereby altering the size distributions.

The remaining possible factors affecting reported particle properties are introduced by the differences in measurement techniques. Uncertainties in number concentrations measured with impactor systems must increase with decreasing diameter. The uncertainty in number concentration ΔN is given by:

$$\Delta N(d) = \frac{\Delta m}{d^3} \tag{4}$$

where Δm is the mass uncertainty which would not depend on particle diameter but on weighing accuracy. For example, consider one 5-μm particle re-entrained off an early impactor stage and attaching to the plate corresponding to 0.5-μm diameter. This would generate an error of 1000 counts in the reported number concentration for 0.5-μm particles, but only an error of 1 count at the 5-μm size. This problem has been observed in impactors and would induce a shift in number concentrations to smaller particles and might explain some of the differences. Similarly, condensation and coagulation phenomena occurring in the sampling train as well as particle size bias introduced by the sampling probe can induce errors in such data.

On the other hand, optical counters inherently operate on a number basis and therefore are less accurate for larger particles which generally occur in fewer quantities. Any random instrument errors are magnified by the few larger particles but statistically averaged out for smaller particles in a given measurement time. However, as previously discussed, the maximum uncertainty expected for the MRSPC is certainly less than an order of magnitude. Additional discussions of the subject have been presented elsewhere (18), and further analysis is underway.

In an attempt to substantiate the applicability of the counter and to evaluate the effects of fuel composition, the particulate emissions of a 15% methanol–85% regular fuel mixture were investigated. Before the measurements were taken, the engine was prepared very thoroughly by running it for about 40 hr. Figure 11 shows the particulate distribution at 424°C in the same observation plane used previously. Addition of methanol seems to have increased the concentration of particles around 1 μm diameter. The difference is nearly indistinguishable at the limits of applicability of the counter.

Figure 11 does not give sufficient information to draw any detailed conclusions on the mechanism by which methanol changes the particle

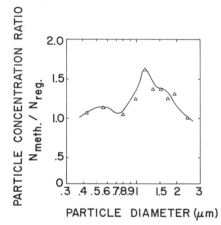

Figure 11.   Effects of methanol addi-
tion (55-mph cruise) 424°C

size distribution. Since neither the large or small particles changed concentration significantly, the origin of the additional particles is not indicated. The effects observed are contrary to our original expectations. Since the addition of methanol leans the mixture, i.e., reduces the equivalence ratio, a decrease of the particulate concentrations was expected. Furthermore, the lower total lead concentration should have furthered this trend as well as the higher temperature at the exit. Our preliminary analysis seems to indicate that the methanol addition causes enhanced coagulation in the nuclei mode with a shift towards the accumulation size range. In general, however, the measurements using the methanol blend fuel seemed to confirm the applicability of the optical counter.

A final observation concerns acceleration effects. It has been frequently observed that during acceleration the amount of particulates emitted is several times greater than that expected for the amount of lead consumed. This can be explained by a sudden removal of the wall deposits caused by a change of the turbulent structure of the flow.

This observation was confirmed in the present study by analyzing the total scattered light signal through a 0–55 mph acceleration. The sudden increase in total scattered light indicates the presence of significantly more particles than at idle. For the short time during acceleration the particle density becomes so great that it is impossible to identify single particles with the present optical configuration. This problem could be handled by reducing the sensitive volume, and an investigation is warranted since significant amounts of particles are emitted during non-steady operation.

Conclusions and Recommendations

The present study represents the first successful application of the optical single-particle measurement concept to in situ, real-time analysis of automobile exhaust gas emissions. Based on revised theoretical considerations of the light scattering and associated error analysis, it was possible to design and build a new generation ratio-type single-particle counter. Elaborate calibration procedures were conducted using particle sizes and materials ranging from submicron polystyrene spheres to micron-size silicon oil droplets of varying density and flow velocity.

Several possible sources of error were identified. The first apparent problem concerns the lack of a unique particle size corresponding to various intensity ratio ranges. This problem is usually concealed during calibration with monodisperse particulates but must be considered when analyzing highly polydisperse combustion-generated aerosols. An effective solution was found in this study by using multiple ratios. Another previously neglected error originates from the dependency of the sensitive scattering volume on the particle size and threshold level. The statistical analysis must be adjusted accordingly.

For proper data acquisition the sampling logic, and subsequently the electronic hardware, was developed that automatically scans the detector output for valid particle signals, completes the ratio operation, and stores the results. A major difficulty in the present context arose from the relatively high speeds of the exhaust gases, and fast ratio circuits had to be developed. Presently, the combination of the counter circuit with a process (mini-) computer system or a larger computer for data acquisition, processing, statistical evaluation, and storage is under development. The advantages of such a system for future studies are obvious.

The MRSPC was applied to the in situ study of automobile exhaust gas emissions. A four-cylinder engine using both leaded gasoline fuel as well as a 15% methanol blend was used to investigate the particulate formation from conventional and alternate fuels. Particulate behavior has gained attention because of the new particulate emission standards. The results indicate considerable deviation from previous data obtained with conventional sampling probe and impactor systems. Also a change in particle size distributions and number density was detected when methanol blends were used as fuel. These effects await further detailed investigations.

In summary, particle size distributions measured at similar conditions using optical and sampling probe–impactor methods are vastly different. No conclusions can be drawn concerning the relative accuracy of these two techniques because of experimental differences. All indications are that the optical counter is operating properly and is applicable and advantageous for in situ measurements. Further experiments comparing the two techniques directly on a common engine will be performed soon to substantiate the particulate formation hypotheses.

Acknowledgments

The authors wish to thank B. Qvale, Laboratoriet for Energiteknik, D.T.H., for his support throughout the project.

Literature Cited

1. Dolan, D. F., Kittelson, D. B., Whitby, K. T., "Measurement of Diesel Exhaust Particle Size Distributions," *ASME Pap.* (1975) **75 WA/APC-5**.
2. Lester, T. W., Wittig, S. L. K., *Symp. (Int.) Combust. Proc. 16th* (1976) (in press).
3. Plassmann, E., "Second Symposium on Low Pollution Power Systems Development," E. Plassmann, Ed., Federal Minister for Research and Technology, TUV Rheinland, Nov. 1974.
4. Bro, K., "Experiments with Methanol, Ethanol, Methane, Hydrogen and Ammonia with Pilot Injection in a Research Diesel Engine," Laboratoriet for Energiteknik Report **RE 75-14**, Technical University of Denmark, 1976.
5. Ganley, J. T., Springer, G. S., "Particulate Formation in Spark Ignition Engine Exhaust Gas," *Fluid Mechanics Laboratory Report* **72-2**, Univ. of Michigan, 1972.
6. Kerker, M., "The Scattering of Light and Other Electromagnetic Radiation," Academic, New York, 1969.
7. Dave, J. V., "Subroutines for Computing the Parameters of the Electromagnetic Radiation Scattered by a Sphere," IBM Report **320-3237**, Palo Alto Scientific Center, Palo Alto, Calif., May 1968.
8. Gebhardt, J., Bol, J., Heinze, W., Letschert, W., *Staubreinhalt. Luft,* English Translation (1970) **30** (6).
9. Hodkinson, J. R., *Appl. Opt.* (1966) **5** (5).
10. Gravatt, C. C., *J. Air Pollut. Control Assoc.* (1973) **23** (12).
11. Phillips, D. T., Wyatt, P. J., *Appl. Opt.* (1973) **11**, 2082.
12. Dalzell, W. H., Sarofim, A. F., *J. Heat Trans (Trans. ASME Ser. C)* (1969) **91**, 100.

13. Gravatt, C. C., "Light Scattering Method for the Characterization of Particulate Matter in Real Time," "Aerosol Measurements," W. A. Cassatt and R. S. Maddock, Eds., *Natl. Bur. Stand. (U.S.) Spec. Publ.* (October 1974) **412,** 21.
14. Gravatt, C. C., Allegrini, Ivo, "A New Light Scattering Method for the Determination of the Size Distribution of Particulate Matter in Air," *Proc. Int. Clean Air Congr. 3rd* (October 1973).
15. Shofner, F. M., Kreikebaum, G., Schmitt, H. W., "In Situ, On Line Measurement of Particulate Size Distribution and Mass Concentration Using Electro-Optical Instruments," *Proc. Ind. Air Pollut. Control Conf., 5th,* (April 1975).
16. Hirleman, E. D., Wittig, S. L. K., "Uncertainties in Particulate Size Distributions Measured with the Ratio-type Single Particle Counter," IEEE-OSA Conference on Laser and Electro-Optical Systems, San Diego, Calif., May 1976.
17. Hirleman, E. D., Wittig, S. L. K., Christiansen, J. V., "Development and Application of an Optical Exhaust Gas Particulate Analyzer," Laboratoriet for Energiteknik Report **RE 76-4,** Technical University of Denmark, 1976.
18. Hirleman, E. D., Wittig, S. L. K., *Symp. (Int.) Combust. Proc. 16th* (1976) (in press).

RECEIVED December 29, 1976. Experiments performed at the Laboratoriet for Energiteknik, The Technical University of Denmark.

Alternate Fuels

15

Microemulsions as Diesel Fuels

G. GILLBERG and S. FRIBERG

Department of Chemistry, University of Missouri–Rolla, Rolla, MO 65401 and
The Swedish Institute for Surface Chemistry, Stockholm, Sweden

Microemulsions with greater than 15 wt % water and cetane numbers larger than 35 show simultaneous large decreases in emitted NO_x and smoke (soot) when used as diesel fuels. Important factors for microemulsion stability are described since these are essential to obtain thermodynamically stable microemulsions. The regions for these are found easily with a systematic knowledge of phase equilibria in surfactant systems. Most surfactants and the microemulsified water had a negative effect on the diesel fuel ignition performance. Conventional cetane number improvers yielded only minor improvements. However, certain emulsifiers function by themselves as cetane number improvers. Relations between emitted NO, smoke, and fuel consumption are given at various injection timings for microemulsions with 10, 20, and 30% water, and the effect in other types of emissions is discussed.

Microemulsions are transparent thermodynamically stable colloidal dispersions containing high amounts of both water and hydrocarbons. The colloidal state is stabilized by a proper balance between a hydrophobic and a hydrophilic surfactant. Initially microemulsions were considered to be different from colloidal solutions (*1, 2, 3, 4, 5*); an opinion that is still held by some (*6*) although it is accepted generally that microemulsions belong to micellar systems (*7, 8, 9, 10*).

This chapter describes the liquid phases in some four-component systems, a treatment that demonstrates the identity between colloidal solutions and microemulsions and gives information about the important factors for their stability.

0-8412-0383-0/78/33-166-221$05.00/0 © American Chemical Society

Water/Oil (W/O) Microemulsions

The basis for the W/O microemulsions is the association between a hydrophobic surfactant (commonly a medium-chain (C_5–C_6) alcohol) and a hydrophilic surfactant (usually a soap in the model systems) such as potassium oleate or an alkyl sulfate. A combination of two such amphiphiles gives ranges of high water solubility capacity according to Figure 1a. In order to show that the three components are equally important, a triangular diagram is often used; the results of Figure 1a then are presented as in Figure 1b. The promotion of solubility of the ionic surfactant from 8 to 45% (Figure 1b) by water is not obtained by inverse micelles. The aggregates in this part of the system contain ion pairs; the notation "micellar systems" which is still used (6) is not correct, simply because no micelles exist in this area.

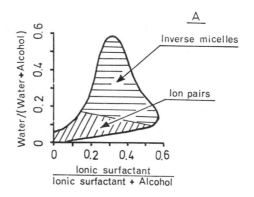

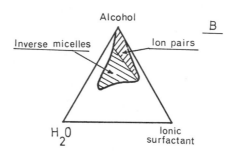

Figure 1.   Solubility of an ionic surfactant in a medium chain length alcohol (C_5, C_6) is promoted by water; ion pairs are formed. The solubility of water is promoted by the presence of the surfactant through the formation of inverse micelles.

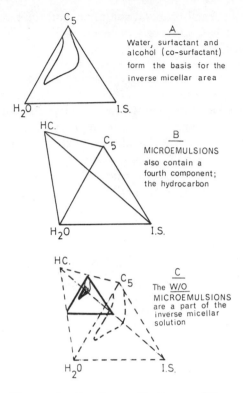

Figure 2.   Starting with the conditions of the three basic components—water, surfactant, and co-surfactant—the micro- emulsions are easily shown to be a part of the inverse micellar area

The enhanced water solubility, on the other hand, is caused by the existence of inverse micelles. The maximum size of these to allow maxi- mum water solubilization depends critically on the alcohol/soap ratio; in Figure 1b maximum water solubility is obtained at a weight ratio of 5:2.

Identical conditions exist if the corresponding solubility region is determined at constant hydrocarbon content in a microemulsion. Such compositions mean that a fourth component is introduced, and a tetra- hedral representation is necessary such as in Figures 2a, b, and c. From this and other diagrams (8, 9, 10) an important conclusion concerning microemulsion conditions may be drawn—the alcohol/soap ratio neces- sary to obtain maximum water solubilization remains identical at different hydrocarbon contents.

This result is important since it saves much experimental labor in order to find optimum compositions for microemulsions. These systems do not differ from normal colloidal solutions, and the term microemulsion is unnecessary. It has, however, been so well established that it will be difficult to change.

Oil/Water (O/W) Microemulsions

These systems have not been investigated as thoroughly as the W/O microemulsions have been. One determination (11) has been reported, and Prince has suggested (6) that the O/W microemulsions exist within a limited oil/emulsifier ratio in a sectorial solubility region emanating from the aqueous corner. This is true only for nonionic systems (12); the combination ionic substance and alcohol gives a more complicated pattern.

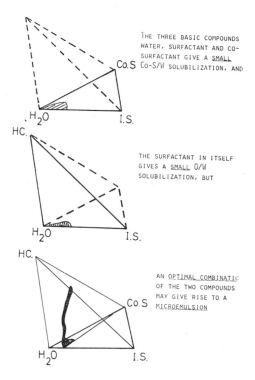

Figure 3. An optimum combination of surfactant and co-surfactant gives micro-emulsions

The most simple representation, found in Figure 3, demonstrates the solubilization dependence on the amount of ionic surfactant dissolved in water. The amount of ionic surfactant in water is the critical condition to obtain O/W microemulsions.

Factors of Importance for the Stability of Microemulsions

The earlier concepts of microemulsion stability stressed a negative interfacial tension and the ratio of interfacial tensions towards the water and oil part of the system, but these are insufficient to explain stability (*13*). The interfacial free energy, the repulsive energy from the compression of the diffuse electric double layer, and the rise of entropy in the dispersion process give contributions comparable with the free energy, and hence, a positive interfacial free energy is permitted.

One important problem connected with the stability of microemulsions is the long-term stability against separation of a liquid crystalline phase. Such a phase is always present in the base diagram, and the liquid crystalline phase which forms a separate phase at equilibrium may be distributed in the form of small spherical particles which are not observed on optical inspection. Prolonged storage, however, may lead to the separation of the liquid crystal, causing turbidity, clogging, and other problems. Phase diagrams such as those described above are a useful tool for clarifying and avoiding problems of this kind since they demonstrate the kind of instability that may be expected at different compositions.

Diesel Fuels

Although several patents concerning the use of microemulsions based on gasoline (*1, 14, 17*) exist, and articles have been published claiming extremely good engine performance with such fuels (*18, 19, 20*), we limit ourselves here to diesel fuels. We believe that microemulsified fuels have their greatest potential when used in the diesel oil engine since the consumption of gasoline is too large for the world production to supply the necessary amounts of emulsifiers for the microemulsified fuels. Furthermore, in contrast to the gasoline engine, no changes in construction of the diesel oil engine will cause its emissions to fulfill the EPA standards.

Many experiments have shown that combustion of petroleum fuels is cleaner and more efficient if small water droplets are dispersed in the oil by ultrasonic, hydrodynamic, or chemical techniques (*21*). The presence of the water leads to a lower generation of nitrogen oxides and to less soot, probably because of lower combustion temperatures and better fuel atomization. The most tested system is a mechanical one for furnaces and ovens produced by the French oil company, ELF (*22*).

The diesel engine requires a fuel with ignition and combustion characteristics opposite from those for the ordinary spark ignition gasoline engine. In the diesel engine an enhanced ignition speed is desired since the fuel is injected, in general, directly into the combustion space.

For its maximum efficiency, the delay period between injection and ignition must be short. The investigations by Valdmanns and Wulfhorst (23) concerning the effect of emulsified water on diesel combustion showed that the increase in ignition delay was large enough to increase fuel consumption, and consequently there were no net reductions in the amount of exhaust pollutants. The emulsion fuels showed no net changes in the amount of NO emitted and only a slight decrease in smoke compared with a standard fuel. Since the economy and benefits of the diesel emulsions were low, the project was abandoned.

Microemulsion Fuels

Ignition delays similar to those observed for the emulsion fuels may be experienced with microemulsion fuels as illustrated by Table I, which shows cetane numbers of microemulsion fuels determined on a CFR engine.

Table I. Cetane Numbers of Microemulsion Fuels

Series	Composition of Fuel (%) (w/w)			Cetane No.
	Diesel Oil	Emulsifiers	Water	
	100	0	0	43
A	80	20	0	37
	76	19	5	34
	68	17	15	31
B	70	30	0	33
	49	21	30	18

The test shows that the emulsifier has a negative effect on the ignition performance and that this effect is further enhanced by adding water. However, an important difference was noted. With microemulsion fuels, the NO_x content of the exhaust gases was reduced substantially while with the emulsion fuels, the detected amount of NO_x increased considerably.

A fuel with a cetane number higher than 40 is generally required for high-speed diesel engines. The cetane number, and thus the ignition of a microemulsion fuel, may be improved using two different approaches. The first one is the conventional addition of cetane-number improvers such as amyl nitrate and kerobrisol MAR. However, at least 10% (w/w) of the improver was needed to restore the cetane number of the microemulsion containing 30% water (Table I) to that of the pure diesel oil.

The second and most promising approach is based on the fact that certain emulsifiers function by themselves as cetane number improvers. Observed cetane numbers of microemulsion fuels produced by this type

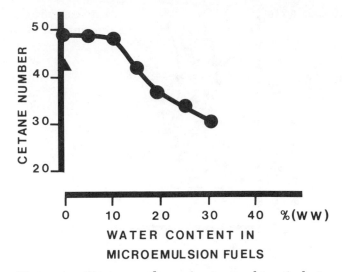

Figure 4. Cetane numbers of microemulsion fuels in which the emulsifiers act as cetane number improvers. (▲) Diesel fuel pure, (●) diesel fuel plus emulsifier and water.

of emulsifier are given in Figure 4. The addition of emulsifier increased the cetane number from 43 to 49, a value which was not reduced by water additions less than 10%.

Tests According to a Limited California Cycle

The engine used was a one-cylinder direct-injection test engine at Saab-Scania, Södertälje, Sweden. During the tests, maximum pressure, fuel consumption, exhaust temperature, CO, CO_2, NO, NO_2, O_2, HC, and smoke were registered at varied injection timings, loads, and speeds. The performance of three microemulsion fuels was compared with a reference fuel of pure diesel oil with the cetane number of 43. The data of the microemulsion fuels are given in Table II.

Table II.   Tested Microemulsion Fuels

Sample No.	Composition (%) (w/w)			Existence Temperature Range (°C)	Viscosity (cP)	Cetane No.
	Diesel	Emulsifiers	Water			
1	63	27	10	0–80	15.7	44–45
2	56	24	20	15–73	14.3	40–42
3	49	21	30	7–65	19.8	35

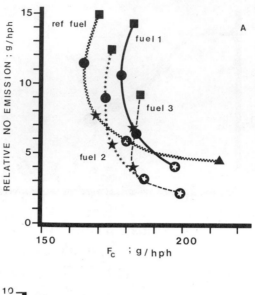

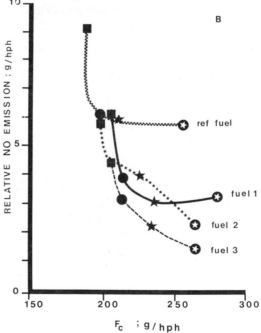

Figure 5.   Relation between the relative NO emission and fuel consumption at various injection timings; ■ 28, ● 23, ★ 18, and ✪ 13 °BTDC. The reference fuel (˙·˙·˙) is a diesel oil with cetane number 43. Fuel 1 (▬▬), fuel 2 (★★★) and fuel 3 (– – –) are microemulsion fuels with 10, 20, and 30% water respectively. (a) Speed 1380 rpm and 100% load (P_e = 8.4 kp/cm²). (b) Speed 2300 rpm and 75% load (P_e = 5.4 kp/cm²).

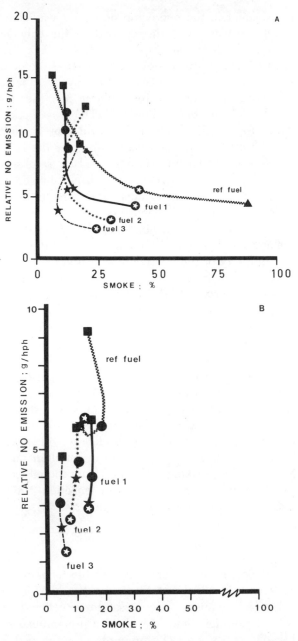

Figure 6. Amounts of emitted smoke and NO at various injection timings; ■
28, ● *23,* ★ *18,* ✪ *13, and* ▲ *8 °BTDC. The reference fuel (* 〰 *) is a diesel*
oil with cetane number 43. Fuel 1 (▬▬ *) , fuel 2 (*★★★*), and fuel 3 (– – –)*
are microemulsion fuels with 10, 20, and 30% water, respectively. (a) Speed,
1380 rpm and 100% load (P_e = 8.4 kp/cm^2*). (b) Speed, 2300 rpm and 75%*
load (P_e = 5.4 kp/cm^2*).*

The water content in the microemulsion fuels was compensated for at injection, and the water was disregarded in the calculations of the fuel consumption. The fuel consumption expressed as g fuel per horse-power-hr increased by approximately 9% for the microemulsion fuels containing 10 and 30% (w/w) of water and by 4% for the fuel with 20% compared with the reference fuel. These increases were approximately independent of injection timing (°BTDC), used speed, and load.

The amount of emitted NO, studied as a function of injection timing, was reduced by 30–65%. The decrease was greater the larger the water content and the higher the speed. Similar trends were observed for NO_2 and smoke. In all cases the microemulsion fuels showed higher levels in oxygen, CO, and HC in the exhaust gas. The critical increase in unburnt hydrocarbons at idling conditions also occurred at a higher injection timing for the microemulsion fuels than for the reference fuel. Changes in cetane number and the composition of the emulsifier seem to improve this property of the microemulsion fuel.

Since an augmented fuel consumption was observed for the micro-emulsion fuels, it is important to determine the net effects, especially for NO and smoke, since all mechanical changes in the engine construction lead to counteractive results for these pollutants. In Figures 5a and 5b the relation between NO emission (expressed in g/hphr) and the fuel consumption is given for two different speeds and loads. Especially at the higher speed of 2300 rpm, it is evident that a positive net effect is obtained by the use of microemulsion fuels. At the lower speed it is only the fuel containing 20 and 30% of water which shows this improvement.

However, if we also take the smoke behavior into account, the advantage of using a microemulsion fuel is clearly evidenced (Figures 6a and 6b). The microemulsion fuel produces a simultaneous decrease in smoke and NO. This is a striking difference from the effects for an ordinary diesel fuel in which the change in injection timing causes a decrease in NO but, at the same time, an immense increase in the soot.

Conclusions

By careful choice of emulsifiers it is possible to provide microemul-sions based on diesel oil which exhibit an ignition performance suitable for high-speed diesel engines. In contrast to emulsion fuels examined previously (23), these microemulsions show a pronounced net benefit in the NO and smoke emissions. The amounts of unburned hydrocarbon and CO in the exhausts do increase but may be reduced by a catalytic afterburner.

The emission changes obtained with the microemulsion fuels parallel those reported for gasoline–methanol fuels to a considerable extent. For these fuels measurements of emitted amounts of polynuclear aromatics (PNA) have also been performed and showed that PNA decreased simultaneously with the decrease in soot and NO. It is therefore most plausible that a similar decrease will be observed for the microemulsion fuel.

The need for emulsifiers and the fact that fuel consumption increased slightly will lead to higher costs if microemulsions are used as diesel fuels. The utilization will therefore be limited to places where the cleaner exhausts regarding soot and NO, aand most probably PNA, will bring considerable savings. Such a case is in mining, where most of the working engines are diesel and in general are already equipped with a catalytic afterburner for CO and HC. More systematic studies of diesel microemulsion fuels are therefore highly motivated.

Literature Cited

1. Kokatnur, V. R., U.S. patent **2,111,100**, 1935.
2. Hoar, T. P., Schulman, H. H., *Nature (London)* (1943) **152**, 102.
3. Schulman, J. H., Stockenius, W., Prince, L. M., *J. Phys. Chem.* (1959) **63**, 1677.
4. Shah, D. O., Hamlin, R. M., *Science* (1971) **171**, 483.
5. Prince, L. M., *J. Colloid Interface Sci.* (1969) **29**, 216.
6. *Ibid.* (1975) **52**, 182.
7. Adamson, A. W., *J. Colloid Interface Sci.* (1969) **29**, 261.
8. Robinson, K., Levine, S., *J. Phys. Chem.* (1969) **76**, 876.
9. Gillberg, G., Lehtinen, H., Friberg, S., *J. Colloid Interface Sci.* (1970) **33**, 40.
10. Shinoda, K., Kunieda, H., *J. Colloid Interface Sci.* (1973) **42**, 381.
11. Kertes, A. S., Jernström, B., Friberg, S., *J. Colloid Interface Sci.* (1975) **52**, 122.
12. Friberg, S., Lapczynska, I., *Prog. Colloid Polymer Sci.* (1975) **56**, 16.
13. Ruckenstein, E., Chi, J. H., *J. Chem. Soc., Faraday Trans.* (1975) **71**, 1690.
14. Robbins, M. L., Schulman, J. H., U.S. patent **3,346,494**, 1967.
15. Brownaweil, D. W., Robbins, M. L., U.S. patent **3,527,581**, 1970.
16. McCoy, F. C., Eckert, G. W., U.S. patent **3,876,391**, 1975.
17. Friberg, S., Lundborg, L., Swedish patent **361,898**, 1974; U.S. patent **3,902,869**, 1975.
18. *Automotive News* (1975) August 11, 13.
19. *Oil Gas J.* (1975) **73**, 60.
20. *Detroit Free Press* (1975) 22 May.
21. Iammartino, N. R., *Chem. Eng.* (1974) Nov. 11, 85.
22. Pariel, J. M., Helion, R., Robic, G., *Rev. Gen. Therm.* (1972) **II**, 979.
23. Valdmanns, E., Wulfhorst, D. E., S.A.E. report **700736**.

RECEIVED December 29, 1976.

16

Evaluation of Oil/Water Emulsions for Gas Turbine Engines

L. J. SPADACCINI and R. PELMAS

United Technologies Research Center, East Hartford, CT 06108

The feasibility of the dispersion fuels concept for application to gas turbine power plants is evaluated from dispersion fuels formulation studies, from the results of single droplet tests directed toward demonstration of the droplet shattering process, and from the results of initial burner tests of dispersion fuels. Results demonstrate the existence of the "microexplosion" phenomenon in single-droplet combustion experiments. Gas turbine combustor tests indicate that fuel emulsification may alter favorably the efficiency of a practical gas turbine combustor without adversely affecting the turbine inlet temperature profile or NO_x, CO, and smoke emissions.

The relatively high cost of light fuel oils continues to affect the economy of this nation adversely because of the large quantity of energy produced by combustion devices that currently require relatively low-viscosity, high-volatility fuels to operate efficiently. The growth rate of power-production-related expenses could be reduced appreciably if such combustion devices could use heavy residual quality oils, and petroleum-based fuels could be conserved if these devices could use coal-derived fuel oils.

However, conversion from light to heavy fuel cannot be accomplished without regard for the deleterious effect the use of a residual quality or coal-derived fuel oil would have on combustion efficiency, pollutant emissions, and engine operational costs. That is, the extent of fuel atomization required for efficient combustion in gas turbine and similar combustion devices cannot be achieved if a high-viscosity heavy fuel oil is simply substituted for a lighter oil. Furthermore, the required atomization cannot be realized easily by modifying the configuration of the fuel injector because injectors compatible with current or currently foreseen combustors do not appear to offer the capability to generate sprays in which most of the mass of a heavy oil would be associated with small size droplets. However, such sprays may be able to be generated by using emulsified fuels in which the dispersed phase (e.g., another fuel or water) exhibits a much higher vapor pressure characteristic than the primary fuel (*1*). Injection of such a dispersion creates droplets in which the primary fuel envelopes a number of droplets of the dispersed phase. As the droplets are heated by combustor inlet air or in the initial stage of combustion, the temperature of the droplet passes through the boiling point level of the dispersed phase, whereupon the volume occupied by the dispersed phase increases by several orders of magnitude. Such volume change induces catastrophic shattering (sometimes referred to as the "microexplosion" phenomenon) of the primary fuel droplet into a number of smaller droplets.

This chapter describes the results of research on evaluating dispersion fuels for application in *gas turbine* engines and includes studies of emulsion formulation, single-*droplet combustion,* and gas turbine combustor tests.

Dispersion Fuel Formulation Studies

Dispersions have been formulated using No. 2 and No. 6 (residual) fuel oils as the continuous phase. Dispersions using No. 6 fuel oil as the continuous phase and water as the dispersed phase were prepared using water-to-fuel oil concentration ratios in the range of 2.5–50%. In all cases the dispersion appeared stable for more than several days. Photomicrographic analyses, such as those shown in Figure 1, indicated that the water was dispersed in the fuel oil in droplets having diameters of approximately 1–5 μ. Photomicrographs of neat (0% water) No. 6 fuel oil are presented for comparison. One notable feature of the residual oil dispersions was that the viscosity increased with increased water concentration. Dispersions containing up to 20% water by weight exhibited a pourability characteristic similar to that of neat No. 6 fuel. However, dispersions containing 50% water by weight were marked by a significant increase in viscosity. Since residual oils are normally preheated prior to atomization, dispersions containing high concentrations of water may require a slightly increased amount of preheating to obtain the equivalent viscosity. Care must be taken that the fuel temperature and pressure are consistent with the vapor pressure characteristics of the dispersed phase. However, as described below, microexplosions have been observed using emulsions with water concentrations considerably lower than 50%.

Dispersions using No. 2 fuel as the continuous phase were also prepared and, in contrast to residual oil dispersions, were inherently unstable, requiring the use of an emulsifying agent(s). A series of stable disper-

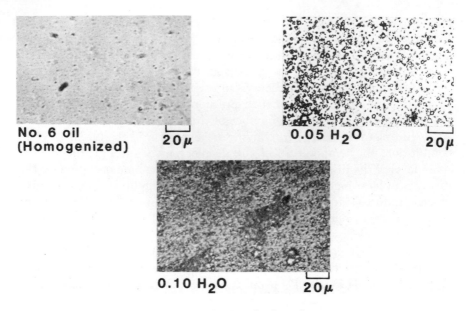

Figure 1.   No. 6 fuel oil emulsions

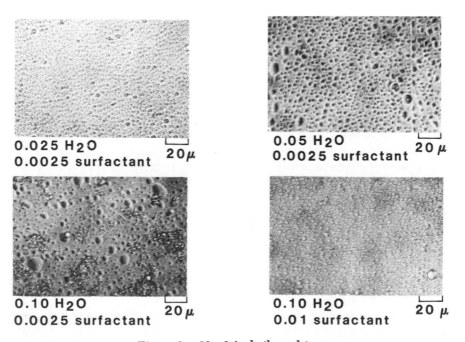

Figure 2.   No. 2 fuel oil emulsions

sions containing up to 10% water in No. 2 fuel oil was prepared by adding approximately ¼–1% surfactant. The surfactant was a sorbitan fatty acid ester formed by a mixture of 75% Span 80 and 25% Tween 85 (Atlas Chemical Ind., Wilmington, DE). Samples of these dispersions are shown in Figure 2. These dispersions were white and cloudy and exhibited no noticeable change in viscosity from the neat No. 2 oil.

Single-Droplet Combustion Studies

Tests comprising the heating and combustion of single droplets of neat No. 6 oil and water/No. 6 oil emulsions were performed to investigate the microexplosion phenomenon. Figure 3 shows the test apparatus

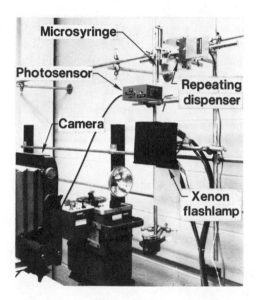

Figure 3.   Single droplet test apparatus

to investigate the heating and combustion of single droplets of an emulsified fuel. Fuel droplets were dispensed from a precision microsyringe, photographed, heated by a helical-shaped xenon flashlamp, and photographed again at the flashtube exit. The flash-lamp was triggered by the output of a photodetector which sensed the falling droplet. Comparisons were made between dispersion fuels and neat fuels undergoing identical test conditions. Droplet heating rates were varied by regulating the energy discharge level of the flashlamp.

Photographs of the heating and combustion of single droplets of neat No. 6 oil and a No. 6 oil–20% water dispersion are shown in the next series of figures. Figure 4 is a progression of photographs of droplets.

exiting the flashlamp after they had been heated by discharges having energies up to 8000 J. Results for droplets of neat No. 6 fuel oil are shown in the upper series of photographs, and the results for droplets of a dispersion of 20% water in No. 6 oil are shown below for comparison. The diameter of the unheated droplets was approximately 1900 μ. In the upper series of photographs, a distinct vapor train emanating from the neat fuel droplet is evident throughout the testing range. The droplet enlarged slightly at 6000 J. The appearance of the neat fuel droplet remained essentially unchanged and showed no signs of shattering. In contrast, the lower series of photographs, which depict the heating of a dispersion fuel, shows the droplet inflating and boiling on its surface at approximately 2000 J. The droplet size increased approximately five to six times its original volume. In addition, the small satellite drops in the photographs indicate that droplet shattering occurred. Additional satellite drops may have gone undetected because the camera lens used had a narrow depth of field. Also, the combustion time required to consume the dispersion fuel droplet was less than was required to burn the neat fuel.

Although these photographs present evidence of droplet shattering, there is no clear indication of microexplosions. Therefore, a high-speed motion picture camera was used to photograph the heating and com-

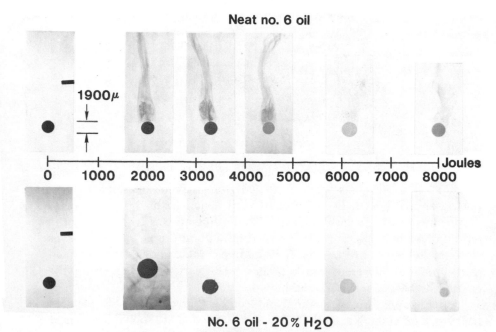

Figure 4.   *Effect of droplet heating*

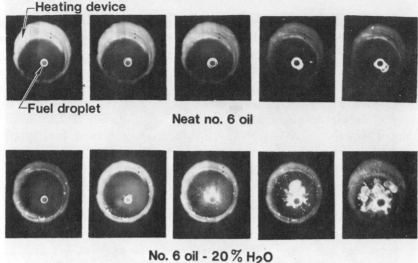

Figure 5.   *Fuel droplet combustion*

bustion of fuel droplets within the flashlamp. The event was viewed indirectly by means of a mirror located on the axis of the flashlamp. The camera was operated at 8000 frames/sec, and the discharge energy of the flashlamp was 1800 J. The results are shown in Figure 5. The upper sequence of photographs, corresponding to neat No. 6 fuel oil, shows the initiation of a flame front and its gradual propagation around the droplet. Again the neat fuel showed no signs of shattering. The results of tests conducted at identical conditions using a dispersion fuel are shown in the lower sequence in Figure 5. Here the film clearly shows the initial stages of droplet boiling and the formation of small satellite droplets. The process continues and results in catastrophic shattering of the primary droplet and combustion of some of the smaller droplets. Finally, a still photograph of the event, representing an integration of the preceding series of frames over the 25-msec flash duration, is shown in Figure 6.

Gas Turbine Burner Studies

Experimental evaluations of fuel combustion and handling characteristics are most conclusive when they are conducted using hardware and test conditions representative of operational combustors. However, full-scale testing of a large industrial gas turbine combustor was beyond the scope of the effort described here. Therefore, the tests were conducted at conditions representative of an FT4 industrial gas turbine

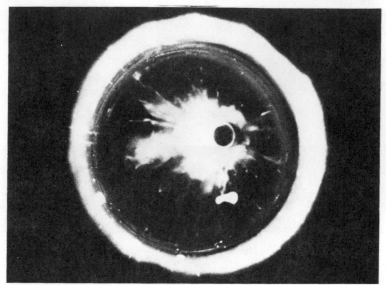

*Figure 6.   Dispersion fuel combustion.   No. 6
oil/20% H_2O.*

engine (40,000 hp) using a smaller but generically similar burner to
forecast changes of a corresponding nature in the operational FT4 combustor. The test combustor consisted of an FT12 burner can which was
modified to operate with inlet air and fuel conditions representative of
an FT4 combustor. No attempt was made to optimize the performance
of the resultant burner assembly. Five homogenized fuels were tested,
each of them based on Redwood 650 oil (*see* Table I) and containing
nominal concentrations of 0, 4, 5, 7.5, and 10% water, respectively. No
emulsifying agent was required. The primary performance parameters
evaluated were combustion efficiency:

$$\eta_c \equiv \frac{\text{actual temperature rise}}{\text{ideal temperature rise}},$$

exhaust temperature pattern factor:

Table I.   Properties of Fuel Oils

	No. 2	*No. 6*	*Redwood 650*
Specific gravity at 298 K	0.845	0.9	0.935
H/C	1.74	1.772	1.22
Wt % N	0.0102	0.106	0.11
Wt % S	0.2960	0.3828	1.10
Viscosity (cs)	5.3 at 298 K	122 at 298 K	200 at 310 K

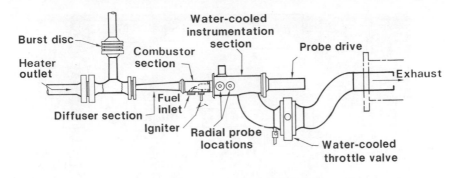

Figure 7.   Combustor test assembly

$$P.F. \equiv \frac{(T_{max} - T_{avg})\ \text{exhaust}}{\text{average temperature rise across burner}},$$

and exhaust smoke. In addition, measurement of the gaseous *pollutant emissions* were included to corroborate other measurement techniques and to elucidate the effects of fuel emulsification.

The combustor test assembly, shown in Figure 7, was comprised of:

(1) An electrical resistance-type heater,

(2) An inlet diffuser,

(3) A cylindrical duct in which the burner can was mounted,

(4) A water-cooled instrumentation section, and

(5) A remotely operated throttle valve, located in the exhaust ducting.

The combustor inlet air pressure was fixed at 10 atm and the temperature at 530 K. The fuel temperature at injection was maintained between 340 and 360 K. The burner was operated at an overall fuel–air mixture ratio of 0.014, corresponding to normal power (40,000 hp) operation of the FT4 gas turbine. Traversing stainless steel, water-cooled probes were

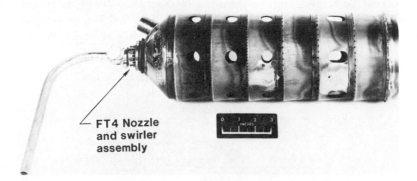

Figure 8.   Modified FT-12 combustor

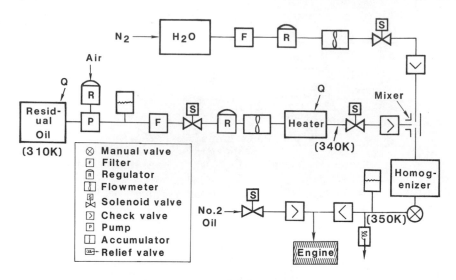

Figure 9.   Fuel delivery system

used to determine the combustor exhaust temperature profiles (from which the combustion efficiencies and pattern factors were deduced), NO_x, CO, HC, and SAE smoke number.

Simulation of the FT4 combustor head end flow characteristics was effected by installing a standard FT4 pressure-atomizing fuel nozzle and air swirler into a FT12 burner can. A photograph of the test combustor is shown in Figure 8. In order to avoid difficulties associated with the use of residual oil, burner start-up and shut-down were effected using No. 2 fuel oil. Therefore, the fuel delivery system (Figure 9), consisted of separate subsystems for supplying No. 2 oil, residual oil, and water. A high-pressure (200-atm) homogenizer, manufactured by Gaulin Corp., was used for continuous on-line fuel emulsification. In addition, heating of the residual oil was required to reduce its viscosity, thereby permitting acceptable flow rates, and to improve homogenization efficiency. However, the fuel temperature was maintained below 370 K to avoid thermal degradation and deposit formation. A positive displacement flow meter

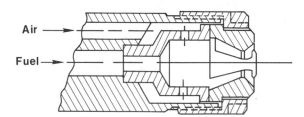

Figure 10.   Air-boost nozzle

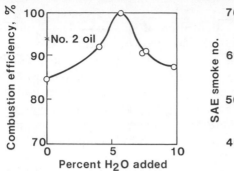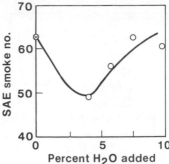

Figure 11.   Combustion efficiency and smoke emissions—pressure-atomizing nozzle.   FT12/FT4 combustor. $T_{air} = 530$ K;   Redwood 650 oil; P $= 10$ atm; $T_{fuel} = 350$ K.

was used to determine the fuel flow rate; the measurement was insensitive to viscosity variation resulting from fuel preheating. In addition, a series of tests was conducted using an air-boost fuel nozzle, shown schematically in Figure 10. This nozzle uses a secondary source of high-pressure air to produce high shearing forces and improved atomization. Boost air at 20 atm and 420 K was supplied to the nozzle at a flow rate equivalent to approximately 2% of the burner airflow rate.

The results of tests conducted using the pressure-atomizing nozzle with emulsified fuels are presented in Figures 11 and 12. They are expressed in terms of the effect of increased water addition on combustion efficiency and pollutant emission rates. The data in Figure 11 indicate that combustion efficiency increased with increased water addition, reaching a maximum at approximately 5% water added, and thereafter decreased, perhaps as a result of local quenching in the primary zone caused by water addition. A 15% increase in efficiency relative to that attained with neat fuel was obtained using a 5% emulsion. The measured efficiency for operation with No. 2 fuel oil is shown for comparison, and differences

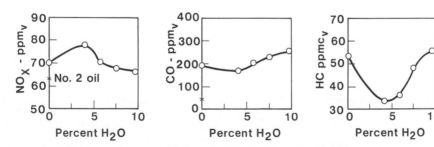

Figure 12.   Gaseous emissions—pressure-atomizing nozzle.   FT12/FT4 combustor. $T_{air} = 530$ K;   Redwood 650 oil; P $= 10$ atm; $T_{fuel} = 350$ K.

may result from less efficient atomization of residual oil and increased time required to vaporize the heavier fractions in the residual oil. Again, no attempt was made to optimize the performance of the FT4/FT12 hybrid burner used in this program. Furthermore, the combustion efficiency of the standard FT4 and FT12 burners using No. 2 fuel oil is approximately 99%. In addition, the temperature distribution at the combustor exit was relatively uniform for all tests in this series, and the pattern factor was always less than 0.2; accepted combustor design criteria require that the pattern factor be less than 0.3.

Smoke emissions, also shown in Figure 11, followed a trend opposite to the combustion efficiency, first decreasing to a minimum value and thereafter increasing. A net reduction in SAE smoke number (2) of 16% was measured for approximately 4% water addition. Thus, the data indicate that fuel emulsification can alter favorably the efficiency of a practical gas turbine combustor without affecting adversely the turbine inlet temperature profile or the smoke emissions. It is speculated that part or all of this improvement may be attributed to improved atomization.

Effects of fuel emulsification on gaseous pollutant emissions are shown in Figure 12, where NO_x, CO, and HC emissions rates are presented as functions of the water concentration in the fuel. The results indicate that NO_x emissions initially increased, achieving a maximum value for 4% water addition, and subsequently decreased, probably from thermal quenching and decreasing reaction rates. A correspondingly opposite trend was observed in both the CO and HC emissions rates.

Although the preceding emissions data are consistent and follow predictable trends, they are not conclusive since they apparently contradict the level of combustion inefficiency deduced from exhaust temperature measurements. It is known that temperatures over 1090 K are required for complete distillation of residual oil. However, since the gas

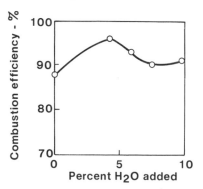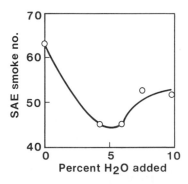

Figure 13. Combustion efficiency and smoke emissions—air-boost nozzle. FT12/FT4 combustor. $T_{air} = 530$ *K; Redwood 650 oil; P =* *10 atm;* $T_{fuel} = 350$ *K.*

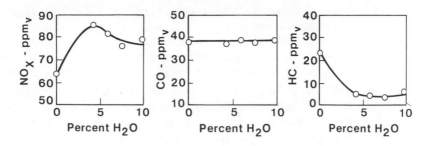

Figure 14. Gaseous emissions—air-boost nozzle. FT12/FT4 combustor. $T_{air} = 530$ K; Redwood 650 oil; P = 10 atm; $T_{fuel} = 350$ K.

sampling and analysis system is limited to operation at 450 K or below, condensation of high-molecular-weight hydrocarbon species can occur and result in loss of sample and, therefore, erroneously low HC emissions measurements. Furthermore, unburned liquid fuel may be present in the exhaust contributing to the error in determination of HC emissions. This fact not withstanding, analysis of the test data confirmed that the gas sampling and analysis techniques conformed to the generally accepted criteria (3) which require that the fuel–air mixture ratio determined from emission measurements agree to within 15% of the value calculated from fuel and airflow measurements.

In order to obtain a more finely atomized fuel spray and simultaneously to alleviate some operational difficulties which stemmed from the accumulation of carbon deposits in the pressure-atomizing nozzle, a series of tests was conducted using an air-boost nozzle. The results, presented in Figures 13 and 14, confirm all of the trends previously established using the pressure-atomizing nozzle. For example, combustion efficiency was maximized and smoke emissions minimized with an emulsion containing 5% water. In addition, the combustion efficiencies obtained were approximately equal to those obtained previously, indicating that perhaps the nozzle spray characteristics were similar or that secondary atomization from droplet shattering was the governing factor. The gaseous pollutant emission rates also followed the previous trends; however, the CO emissions levels were reduced significantly with air-boost atomization. This may result from the introduction of boost air which effects an initial degree of air–fuel premixing and also decreases the local fuel–air mixture ratios in the primary zone.

Conclusions and Recommendations

The principal conclusions derived from this work are:

(1) Emulsified fuels can undergo secondary atomization in combustion systems as a result of heating and expansion of the internal (dispersed) phase.

(2) Fuel emulsification may alter favorably the efficiency of a practical gas turbine combustor without affecting adversely the turbine inlet temperature profile or the pollutant emissions rates.

The results of this investigation indicate that additional studies are necessary to evaluate fully the extent to which fuel emulsification can, within acceptable air pollutant emission limits, broaden the range of fuels compatible with current combustion devices. Furthermore, presuming that emulsification does offer that benefit, consideration must be given to defining procedures for the design of combustors using such fuels, defining specifications to modify installations using emulsified fuels, and assessing modification and operational costs for such installations.

Literature Cited

1. Ivanov, V. M., Nefedov, P. I., "Experimental Investigation of the Combustion Process of Natural and Emulsified Fuels," Trudy Instituta Goryachikh Iskopayemykh, 19 (Russian), NASA Tech. Translation TTF-258, 1965.
2. "Aircraft Gas Turbine Engine Exhaust Smoke Measurement," Aerospace Recommended Practice 1179, SAE, 1970.
3. "Procedure for the Continuous Sampling and Measurement of Gaseous Emissions from Aircraft Turbine Engines," Aerospace Recommended Practice 1256, SAE, 1971.

RECEIVED December 29, 1976.

Feasibility of Methanol/Gasoline Blends for Automotive Use

R. T. JOHNSON and R. K. RILEY

Mechanical Engineering Department, University of Missouri—Rolla, Rolla, MO 65401

A three-phase test program evaluated methanol/gasoline blends as possible automotive fuels. Octane rating of the blends demonstrated that methanol substantially increased the research octane number (RON) and had very little effect on the motor octane number (MON). Single cylinder engine tests indicated that 10% methanol/gasoline blends did not substantially alter power, emissions, or fuel economy of the engine. The vehicle portion of the test program included an ordinary spark ignition engine vehicle calibrated to meet 1974 standards and a prechamber, stratified-charge engine vehicle. HC and CO emissions were reduced using a 10% methanol blend in the 1974 calibration vehicle. Emissions and fuel economy were not substantially altered when a 10% methanol blend was tested in the stratified-charge engine vehicle.

The use of alcohol as an internal combustion engine fuel is a concept that has been in and out of vogue since the early 1900s. These cycles of interest began in the early 1920s and seemed to peak about every 10 years until the present. Bolt (*1*) presents an interesting and informative discussion of these recurring efforts through the early 1960s. The incentives for these examinations of alcohols as motor fuels ranged from testing them as antiknock agents to evaluating ways to use surplus grain commodities by making ethyl alcohol.

The current interest in alcohols as motor fuels is rooted in the search for alternate fuels to replace our suddenly limited petroleum-based fuels. The fact that methyl alcohol (methanol) can be produced from a variety of sources, including coal and garbage, has focused considerable attention on this material as a possible alternate fuel. The fact that the tech-

nology exists today to construct large scale plants to manufacture methanol from coal or waste products adds to the reasons for thoroughly considering this material as a liquid fuel alternative to gasoline.

One major problem in considering methanol for an automotive fuel
is that it has a significantly different stoichiometric air–fuel ratio than
gasoline. This means that pure methanol would not operate in an unmodified internal combustion engine calibrated to use gasoline. However,
this problem is largely overcome if methanol is used with gasoline in
small percentages. For methanol/gasoline blends of 15% methanol or
less, the fuel stoichiometry is such that most internal combustion engines
will operate. For this type of use, methanol could be considered a fuel
extender. Some authors (2, 3) have indicated that methanol/gasoline
blends not only improve fuel economy and reduce exhaust emissions but
reduce engine knock problems as well. A fuel blending material with
these properties certainly deserves additional examination.

This chapter presents a relatively unbiased evaluation of the advantages and disadvantages of methanol/gasoline blends as fuels for spark
ignition engines. The emphasis is on the use of these blends as possible
fuels for the existing and near-future U. S. auto population. The evaluation is divided into three basic areas:

(1) Evaluation of the octane-improving characteristics of methanol
blended with several gasoline base stocks.

Table I.   General

Characteristic	Base Fuel A
General description	summer blend of regular grade gasoline without TEL
RON	81.1
MON	75.5
Nominal HC distribution (%)	
olefins	1.0
aromatics	31.0
saturates	68.0
ΔN (RON)	30.35
ΔN (MON)	10.65
K (RON)	0.02759
K (MON)	0.03743

(2) Fuel research engine evaluation of the performance and emissions of methanol/gasoline blends at equivalent operating conditions to determine any benefits from the slightly altered fuel chemistry.

(3) Vehicle tests to evaluate the performance, emissions, and fuel economy of methanol/gasoline blends in current and near-future vehicles.

Octane Characterization of Methanol/Gasoline Blends

Various authors (2, 3, 4, 5) have cited the antiknock qualities of methanol blended with unleaded gasoline. Unfortunately, comprehensive data on the performance of methanol as an octane improver for unleaded gasoline have not been readily available in the open literature. This is particularly true for information concerning the more severe motor octane rating of blends. A program to evaluate the octane characteristics of blends of methanol and unleaded gasoline was therefore initiated. The results of this program are summarized below. A more detailed presentation of the program and results is given in Ref. 6.

Octane Test Program. In planning the test program for the methanol/gasoline blends, two objectives were established. The first objective was the octane rating of methanol blended with unleaded gasolines. Unleaded fuels were chosen since any large-scale use of methanol blended with gasoline is several years away, and it is anticipated that most cars will then require unleaded fuel.

Base Fuel Characteristics

Base Fuel B	Base Fuel C	Base Fuel D
summer blend of regular un- leaded gasoline commercially available	summer blend of premium type unleaded gaso- line for 1975 and later ve- hicles	Indolene (14) unleaded, ref- erence fuel specified for use in vehicle emission certi- fication
89.9	95.8	98.1
82.1	87.1	87.6
12–15	1.0	10
20–25	27.5	35
68–60	71.5	55
20.85	14.05	12.00
4.10	1.96	0.66
0.02583	0.03426	0.02896
0.02919	0.0251	0.0345

A second objective of this study was to develop a convenient, but not necessarily rigorous, mathematical description of the octane number of blend (ONB) as a function of the base fuel and the volume percent methanol. For this reason, tests were planned for methanol percentages greater than 50% even though these mixtures were not considered viable fuels. Complete information for research and motor ratings of all base fuels with 0–100% methanol was desired.

Four unleaded fuels were selected for the octane evaluation program. The fuels ranged from a regular grade of leaded gasoline without the tetraethyl lead to Indolene high octane, the unleaded fuel used for emissions certification. Indolene is the trade name for the test fuel manufactured by Standard Oil Co. to meet the specification called out in Ref. 7. Some of the base fuel properties are given in Table I.

In general, the procedures used to determine the octane characteristics of methanol/gasoline blends are those specified in ASTM D 2699 (8) for research octane number (RON) and ASTM D 2700 (8) for motor octane numbers (MON). Some minor equipment changes were made to maintain proper engine operating conditions when testing blends containing a large percentage of methanol. The octane ratings of 100% methanol were obtained using a modified rating procedure similar to that used for benzene. A more detailed description of the exact procedure is contained in Ref. 9.

Results of Octane Test Program. Figures 1–4 show the octane characteristics of the blends formed with methanol and each base fuel. For the information presented, % methanol is the vol % of methanol in the blend based on the individual constituent volumes before mixing. This is a seemingly minor but important point since actual mixing does not produce a volume equal to the sum of the volumes of the two constituents.

The data show that methanol improved the RON for all base fuels. The effect of methanol on the MON is not nearly as pronounced as on RON. Only base fuels with a MON less than 83 showed any significant octane number increase. Base fuels with MONs above 83 showed substantial improvement or even loss in MON with the addition of methanol.

Mathematical Description of Results. A convenient mathematical relationship was arbitrarily selected to help to describe the effect of methanol on the fuel octane rating. The equation used was:

$$ONB = N_f + \Delta N(1 - e^{-Kx})$$

where ONB = octane number of blend; x = vol % methanol in blend; N_f = octane number of base fuel; ΔN = octane increment from addition of methanol (ΔN > maximum octane number increase from methanol); and K = base fuel response factor. N_f, ΔN, and K are the parameters and,

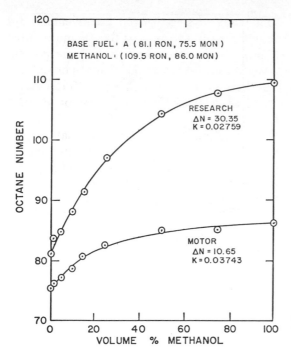

BASE FUEL: A (81.1 RON, 75.5 MON)
METHANOL: (109.5 RON, 86.0 MON)

RESEARCH
$\Delta N = 30.35$
$K = 0.02759$

MOTOR
$\Delta N = 10.65$
$K = 0.03743$

Figure 1.   Blend octane characteristics—base fuel A and methanol

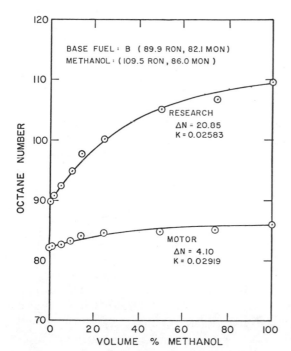

BASE FUEL: B (89.9 RON, 82.1 MON)
METHANOL: (109.5 RON, 86.0 MON)

RESEARCH
$\Delta N = 20.85$
$K = 0.02583$

MOTOR
$\Delta N = 4.10$
$K = 0.02919$

Figure 2.   Blend octane characteristics—base fuel B and methanol

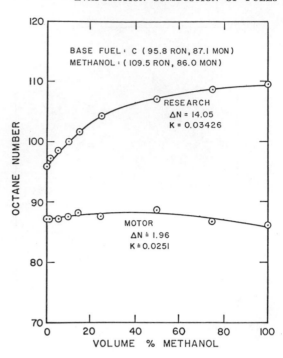

Figure 3.   Blend octane characteristics—base fuel C and methanol

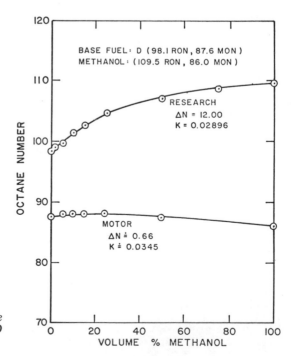

Figure 4.   Blend octane characteristics—base fuel D and methanol

in keeping with the desire for a physical interpretation, require additional description. N_t can be either a research number or a motor number depending on whether the equation is being used to calculate research or motor octane numbers. The parameter ΔN is fixed for any given base fuel and indicates how effective methanol will be in raising the octane of the base fuel. Large values of ΔN correspond to large possible octane increases from blending with methanol. K describes the effect of small concentrations of methanol on the ONB. K is fixed for a given fuel and is a relative term that must be used in conjunction with ΔN. In comparing the effects of methanol used with different base fuels, K indicates whether the major octane increase will be at low or high methanol con-

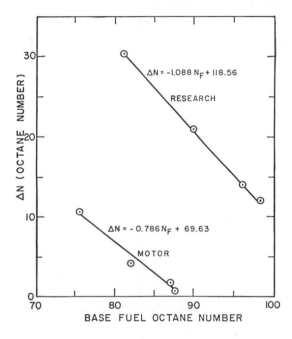

Figure 5.   *Correlation between octane increment and base fuel octane number*

centrations. Larger values of K indicate that a larger portion of ΔN will be obtained for a given concentration of methanol. Thus, if two base fuels have the same value for ΔN, the fuel with the larger K will have the higher blend octane number for a given menthanol concentration. Conversely, if two base fuels have the same value for the parameter K, the fuel with the larger ΔN will have the higher blend octane number for a given methanol concentration.

A least square error technique was used to determine values for the parameters ΔN and K from the experimental data. Examination of these parameters and base fuel properties yielded one significant correlation between ΔN and N_t. Figure 5 illustrates this straight line relationship.

Single Cylinder Engine Tests

The basic purpose of the single cylinder engine test program was to characterize how engine parameters such as speed, air–fuel mixture, and spark advance affect the emissions, fuel economy, and performance of methanol/gasoline blends. In order to examine the effects from the addition of methanol to gasoline, direct comparisons were planned where identical tests would be performed on the base fuel and the methanol–base fuel blend. When reasonable limits were applied to the data that could be taken, it was decided to examine only blends of 10% by volume of methanol with the base fuels used. This decision was based on several facts. The results of the octane testing program indicated that the greatest relative effect of any octane improvement from the addition of methanol would be obtained with methanol concentrations of 10% or less. The phase separation problem between gasoline and methanol in the presence of small amounts of water indicated that feasible mixtures of more than 10% by volume of methanol with gasoline required additional stabilizing chemicals.

Engine Operating Conditions and Instrumentation. The single cylinder engine used for this test program was a Waukesha variable compression ratio CFR engine designed primarily for octane rating of fuels. For the tests performed in this program the compression ratio was set at 6:1 based on the nominal octane of the fuels to be tested and the desire to operate over a wide range of ignition timing without fuel detonation.

Normal spark ignition engines operate over a range of air–fuel mixtures. Maximum power is obtained under fuel–rich conditions while best economy is obtained under fuel-lean conditions. To make comparisons between fuels, it is necessary to define the air–fuel equivalence ratio, ϕ_{AF}. ϕ_{AF} is the ratio of the actual air–fuel ratio to the stoichiometric air–fuel ratio. Thus, ϕ_{AF} is less than one for rich operation and greater than one for lean operation. Equivalence ratios between 0.9 and 1.2 were examined in this study.

Several other variables were examined in addition to the equivalence ratio. The engine operating speed was varied from 600 to 1800 rpm. Ignition spark timing was varied from MBT (minimum advance for best torque) to MBT − 15 degrees. Other variables were fixed with the aid of various humidity and temperature control systems. A schematic of the engine apparatus is shown in Figure 6. A substantially more detailed explanation of the test engine, operating conditions, and test procedures is given in Ref. 6.

Emissions-measuring instruments used in this program consisted of non-dispersive infrared (NDIR) analyzers for CO, CO_2, and NO. A modified flame ionization detector (FID) was used to measure HC. Modifications insured that the instrument would detect the unburned methanol in the exhaust. Additional information concerning the analysis system is provided in Refs. *6* and *9*.

Results. All emissions data were reduced to a mass basis using a carbon balance technique developed by Stivender (*10*). These data were then put on a specific basis by dividing the mass of emissions by the energy produced by the engine in indicated horsepower hours (ikw-hr). Thus, the emissions are reported as grams per indicated horsepower hour (g/ikw-hr) of the particular species. The use of indicated power (ihp or ikw) minimizes the effect of the large frictional losses in the CFR engine on the specific emissions and makes the results more comparable with commercial engines.

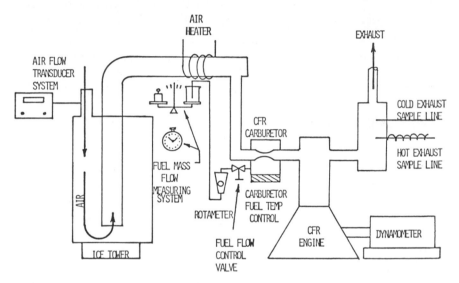

Figure 6.   Schematic of single cylinder engine test apparatus

Data were taken for four different fuels over the complete range of test conditions. The fuels were base fuels C and D and each of these base fuels plus 10% methanol. As noted in Table I, base fuel D is unleaded Indolene, the reference fuel specified for federal emissions tests.

In order to easily interpret the experimental information, the data are presented in graphical form. Significant changes produced by the addition of methanol or any of the engine variables are obvious in this direct presentation.

Five dependent variables were selected for graphical presentation:

(1) Indicated specific hydrocarbons, ISHC [g/ihp-hr (g/kw-hr)];

(2) Indicated specific carbon monoxide, ISCO [g/ihp-hr (g/kw-hr)];

(3) Indicated specific nitric oxide, ISNO [g/ihp-hr (g/kw-hr)];

(4) Indicated specific fuel consumption, ISFC [lb/ihp-hr (kg/kw-hr)];

(5) Indicated thermal efficiency (η_i).

The independent variable used for all graphs was the air–fuel equivalence ratio ϕ_{AF}. Data were taken for each of the four engine speeds tested. However, only the information for 600 rpm is presented here. Generally, data for the other speeds are similar to those presented for 600 rpm. A complete set of information is contained in Ref. 9. Each graph also shows the effect of spark retard and of the addition of 10% methanol through the use of shaded and unshaded symbols.

Examination of the data indicated that, generally, there was very little difference between the results for base fuel C and/or Indolene (base fuel D). In order to avoid redundant information, only the data for the Indolene fuel system are presented in this paper. Where any significant differences between the two fuel systems tested occurred, a comment or description is included. The data for the Indolene fuel system were chosen for presentation because this fuel is a national standard, and its composition and characteristics are well known. Figures 7–11 are the graphical results for the Indolene fuel system. Each figure specifies the test conditions and values of the different parameters that were varied. Only trend lines have been included on the graphs since a line for each set of data points would lead to a confusing number of lines on each figure.

Figure 7 reveals that the addition of 10% methanol to Indolene does not substantially alter the HC emissions. The effect of spark retard is

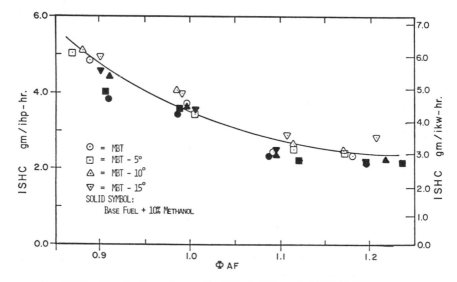

Figure 7.   Indicated specific HC, Indolene base fuel, 600 rpm

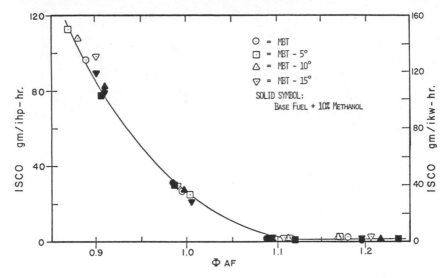

Figure 8.   Indicated specific CO, Indolene base fuel, 600 rpm

relatively small on the HC emissions of both the clear and blended fuels. At low engine speed the retarded spark causes some increase in the ISHC emissions. As the engine speed is increased, the effect of spark retard is reduced, and the ISHC emissions generally decrease. There were slight differences in the changes in the ISHC emissions caused by adding methanol between Indolene and base fuel C. The differences were small and most obvious at low engine speed and low equivalence ratio. Under these conditions, the Indolene base fuel blend showed no change to a slight reduction in the ISHC emissions. Methanol blended with base fuel C showed no change to a slight increase in the ISHC emissions. At higher engine speeds and equivalence ratios there was very little difference in the ISHC emissions for all the fuels.

As expected, adding 10% methanol to the base fuel had little effect on the ISCO emissions. Figure 8 clearly demonstrates that the ISCO emissions are almost entirely a function of the equivalence ratio. Even the spark retard and engine speed have little effect on these emissions.

Nitric oxide emissions are influenced more by changes in equivalence ratio, spark retard, and the addition of methanol to the base fuel than either of the other measured emissions. In order to distinguish clearly between the effects of engine parameters and the addition of methanol on the ISNO emissions, the changes from the engine parameters alone are discussed first. Figure 9 demonstrates clearly the expected characteristic relationship between NO emissions and air–fuel equivalence ratio. The ISNO emissions show a maximum value at an air–fuel equiva-

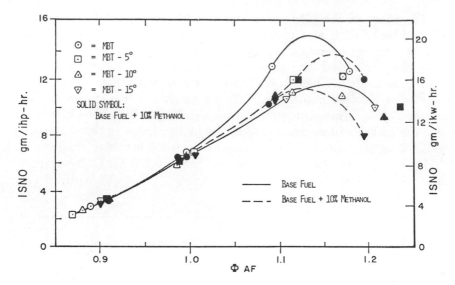

Figure 9.   Indicated specific NO, Indolene base fuel, 600 rpm

lence ratio on the lean side of stoichiometric. Spark retard has a pronounced effect in reducing the ISNO emissions, particularly at equivalence ratios near that for peak ISNO emissions. These engine parameter effects on the ISNO emissions are typical and of the magnitude expected for the single cylinder engine.

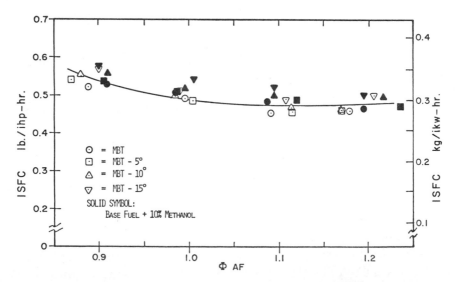

Figure 10.   Indicated specific fuel consumption, Indolene base fuel, 600 rpm

The effect on the ISNO emissions of adding 10% methanol to the Indolene base fuel is also shown in Figure 9. Some differences can be noted at different operating conditions. Generally, the maximum ISNO values are unchanged or somewhat lower for the methanol–Indolene blend. The methanol–base fuel C blend demonstrated slightly different behavior in that the maximum ISNO for the blend was generally the same or slightly greater than for the base fuel. For both base fuel systems, the addition of methanol caused slight alterations in the shapes of the ISNO vs. ϕ_{AF} curves. There was no clear trend for these changes.

Figure 10 illustrates the relationships between the ISFC, the engine parameters, and fuel alterations. As anticipated, increasing equivalence

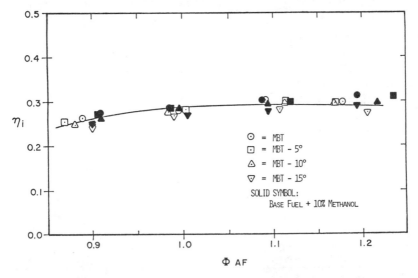

Figure 11.   Indicated thermal efficiency, Indolene base fuel, 600 rpm

ratio reduced the ISFC. Spark retard produced equivalent increases in the ISFC at virtually all operating conditions. The addition of 10% methanol increased the ISFC for all operating conditions. This increase was expected because of the reduced energy content of the methanol blend fuel.

Figure 11 displays the energy consumption characteristics of the clear and blended fuels on a thermal efficiency basis. From this information it is apparent that, at equivalent operating conditions, there is essentially no difference in engine efficiency caused by the addition of methanol. The figures also indicate that if the engine can be operated at greater (leaner) equivalence ratios, some increases in thermal efficiency can be expected.

Vehicle Test Program

The vehicle test program to evaluate methanol/gasoline blends as vehicle fuels was divided into two separate phases. Phase I was a chassis dynamometer evaluation of the emissions and fuel economy of a small displacement engine and drive system that met 1974 emissions standards. Phase II was a somewhat different program to evaluate the emissions, fuel economy, and driveability of methanol/gasoline blends in a stratified-charge engine vehicle. Since there were some major differences in the evaluation programs, each is discussed separately.

Phase I—Chassis Dynamometer Tests. A 1.3-L engine equipped with a four-speed automatic transmission and complete drive system was calibrated to meet 1974 emissions standards using Indolene reference fuel. This standardized apparatus was then used in conjunction with the federal emissions test procedure to evaluate the performance of methanol/gasoline blends. Results of the single cylinder test program and preliminary evaluations with the chassis dynamometer system indicated that blends should be limited to 10% methanol or less to avoid driveability and other problems. A complete discussion of the apparatus and test procedures is given in Ref. 9.

The results of the chassis dynamometer test program are presented in Figures 12 and 13. For the base fuels tested, the most notable change is the decrease in CO emissions. A nominal reduction of 30% is obtained by adding 10% methanol to the base unleaded gasoline. This reduction has also been noted by Wigg and Lunt (*11*) and by Brinkman et al. (*12*).

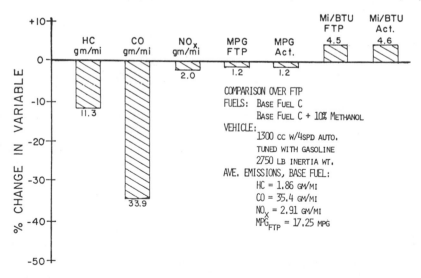

Figure 12. Changes in FTP results for 10% methanol added to base fuel C

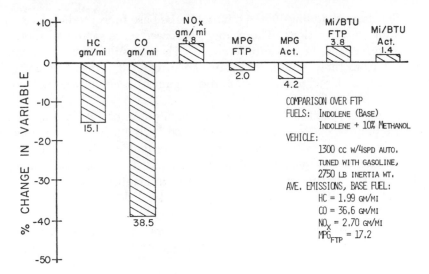

Figure 13.   Changes in FTP results for 10% methanol added to Indo-lene

Since CO emissions are almost entirely a function of the mixture stoichiometry, the reduction in CO emissions can be attributed to the effective leaning of the air–fuel mixture by the addition of 10% methanol.

Addition of 10% methanol appeared to reduce slightly the HC emissions. However, examination of the data for each cycle of the procedure revealed that the reduction occurred during the first few cycles of the cold transient portion of the test. After the engine was fully warmed up, there was virtually no difference in the HC emissions between the fuels containing methanol and the base fuels. This indicates that the leaning effect produced by adding methanol produces a significant reduction in HC emissions during cold operation with an enriched mixture. Once the engine is warmed up and normal air–fuel mixtures are obtained, the added methanol has little effect on the HC emissions. If the automatic mixture control were altered to compensate for the leaning effect of the methanol, the apparent reduction in HC emissions would be greatly reduced. Brinkman et al. (78) tested a number of different vehicles representing the current automobile population and found no significant change in the HC emissions when 10% methanol was added to the base fuel. These authors also noted that mixture enrichment during cycle 1 and 19 of the federal test procedure could cause test results that did not truly indicate how a vehicle fueled with a methanol/gasoline blend would behave under average operating conditions. Wigg and Lunt (11) also note that the change in HC emissions produced by adding methanol is small for late-model vehicles and is of little real significance.

The effect of adding 10% methanol to the base fuels produced varied results for NO_x emissions. The changes observed were small and showed both positive and negative differences. These results are not greatly different from those described by Brinkman et al. (12) and Wigg and Lunt (11), and the magnitude of the changes observed was not considered to be significant.

Changes in the fuel economy from the addition of 10% methanol to the base fuel were essentially those indicated by the leaner stoichiometry of the blends. Generally, the miles traveled per gallon of blend decreased from that of the base fuel. On the other hand, the miles traveled per million BTU's of energy increased slightly, indicating a slight increase in the efficiency of the engine because of the blended fuel. These changes agree well with the results of Wigg and Lunt (11) and of Brinkman et al. (12) and can be attributed to the leaner operation of the engine. Fuel economy was based upon actual fuel consumption and was not computed from emissions data.

Although the driveability of the test fuels was not evaluated objectively, subjective observations were recorded for each test run. These subjective observations can be summarized as follows:

1. Generally, adding 10% methanol to the base fuel degrades the vehicle driveability. This effect is particularly noticeable when attempting to accelerate with a cold engine. Considerable sag and hesitation were evident during the first few accelerations of the driving schedule.

2. After the vehicle was fully warmed up, there were no noticeable differences in performance between the base fuels and the base fuels plus 10% methanol.

3. When operating on fuels containing 10% methanol, the engine idle speed was reduced, and the engine did not idle as smoothly as with the base fuel alone.

Other than these differences, little change in vehicle driveability between the methanol blends and the base fuels was noted.

Phase II, Stratified-Charge Engine Vehicle Program. A combined interest in methanol/gasoline blend fuels and stratified-charge engine technology led to the examination of methanol/gasoline blends as fuels for stratified-charge engine vehicles. A 1975 Honda vehicle with a 1.5-L compound vortex controlled combustion (CVCC) engine was available for tests. In addition to the chassis dynamometer test procedures, the road operation of the vehicle with various fuel blends was evaluated. A complete discussion of the vehicle and the different test procedures is given in Ref. 13.

The road test portion of the stratified-charge engine program was completed first. Blend concentrations ranging up to 40% methanol were evaluated in these tests to determine the possible operating limits of the blends in the CVCC vehicle. Leaded base fuels were used for these vehicle tests since the manufacturer recommended that leaded fuels be used at least occasionally in the vehicle. Results reported here are for a commercial summer blend leaded fuel and Indolene-30, a high octane leaded reference fuel specified in Ref. *14*. Generally speaking, the properties of the Indolene-30 are similar to those of Indolene unleaded.

Figure 14 illustrates how the addition of increasing concentrations of methanol influence the road driveability of the CVCC vehicle. For the driving course used in this study and the rating system used to assign demerits, a vehicle with extremely poor driveability will rate 150–200

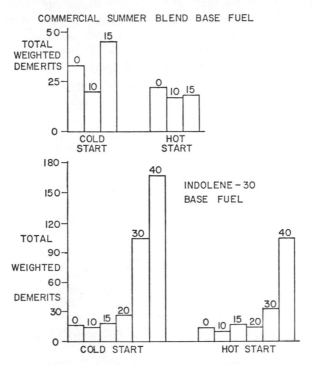

Figure 14. Summer driveability—total weighted demerits. Numbers over bars = methanol.

total weighted demerits (TWD). A vehicle with good driveability will rate less than 20 TWD. This is a smaller range of demerits than normally reported for CRC type driveability tests. This variation is not considered detrimental since all tests are made on a comparative basis, and the objective of the tests is not to establish the absolute driveability of the vehicle but to compare different fuels. Figure 14 demonstrates that adding small concentrations of methanol to the two base fuels does not produce major changes in the driveability of the vehicle. In fact, the 10% methanol blends showed a slight improvement in the vehicle driveability over that for the base fuel alone. A blend of 15% methanol with the base fuels caused deterioration of the driveability to a level roughly comparable with that of the base fuel. Generally the addition of small concentrations of methanol did not substantially alter the TWD values for the Indolene-30 fuel system. For methanol concentrations above 20%, the TWD values increase rapidly, indicating substantial deterioration in the vehicle driveability. These results were expected based on the fuel metering characteristics of the CVCC engine. The small blend concentrations of methanol produced changes in the fuel stoichiometry of only 5–10%. A leaning of the mixture by 5–10% in the prechamber will do very little to alter the ignition characteristics of the rich prechamber mixture. This is true even at the higher concentrations of methanol because of the very rich prechamber mixture. However, the driveability deteriorates at the higher concentrations because of the excessively lean mixture in the main chamber, which causes the overall A/F ratio to be excessively lean. This leads to loss of power and other lean operation symptoms.

Figure 15 illustrates how the addition of methanol to the base fuel affects the acceleration times between 25 and 55 mph (33.0 and 88.5 kph) for various operating conditions of the vehicle. For blends of up to 15% methanol there is little effect upon the wide open throttle acceleration times. Under part-throttle operation at 5 in. Hg (16.9 kpa) and crowd conditions (no opening of secondary throttle), the addition of methanol

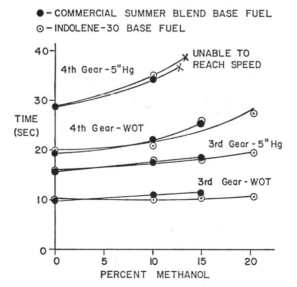

Figure   15.   Summer   driveability—acceleration times, 25–55 mph

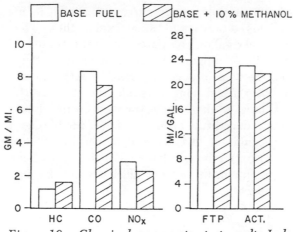

*Figure 16.   Chassis dynamometer test results Indo-
lene-30 base fuel*

to the base fuel increased the acceleration times. This reduction in response to throttle input is usually termed "stretchiness." Wigg and Lunt
(*11*) have indicated that this increased "stretchiness" is a typical characteristic of part-throttle operation with methanol/gasoline blends in normal spark ignition engine vehicles. This characteristic could be anticipated in that road load energy requirements will demand a greater volume of the lower heating value methanol/gasoline blend. A greater
throttle opening will be required to provide the required increase in fuel
volume.

Given limited resources, the information developed in the single-
cylinder engine and chassis dynamometer test programs and the results
of the road test program with the stratified-charge engine vehicle, a
blend of 10% methanol and gasoline was selected as most appropriate
for evaluating emissions and fuel economy. All tests were performed
using the Indolene-30 base fuel. The bar graphs in Figure 16 illustrate
the effect of adding 10% methanol to the Indolene-30 base fuel. There
appear to be some changes produced by the addition of methanol to
this base fuel. However, the only changes that have statistical significance (95% confidence level) are those for HC and NO_x. The increase
in HC emissions can be traced directly to increased starting problems
with the methanol blends, particularly during the hot start portion of
the test. This problem is well documented with the modal analysis technique used for the test program. The slight reduction in the NO_x emissions is attributed to the leaning of the air–fuel mixture caused by adding
methanol. Although the change in CO emissions is not statistically significant, it is typical of results expected from additional leaning of the
air–fuel mixture.

The two fuel consumption measurements shown in Figure 16 (and Figures 12 and 13) deserve additional explanation. The mpg (FTP) value is the miles per gallon computed using the statute 7.5 miles (12.1 km) specified in the computation of emissions from the urban driving schedule. Since the vehicle does not travel exactly 7.5 miles (12.1 km) when operating over this driving schedule, the mpg (ACT) is computed using the actual mileage the vehicle traveled on the dynamometer rolls. Even though the reduced fuel economy shown for the methanol blend is not statistically significant, it is typical of changes that have been reported (*6, 11, 12*) for methanol blends. For these small changes in volumetric fuel consumption, the energy efficiency can be considered to be substantially unaffected.

Conclusions

The results of the octane evaluation program support current opinion and data for research octane ratings; methanol can produce a significant increase in the RON of unleaded gasoline, particularly if the base fuel RON is less than 90. Generally, the higher the RON of the base fuel, the lower the antiknock improvement achieved by adding methanol. The motor ratings showed significantly smaller octane increases than those found for the research method. Even those base fuels with a relatively low MON did not show really significant increases in MON when blended with methanol. Base fuel with a MON greater than 83 gave virtually no change in MON when blended with methanol.

The octane behavior of the blend casts serious doubt on the use of methanol to improve the MON of unleaded gasolines. However, these data also indicate that methanol is not likely to degrade the MON rating of an unleaded gasoline to which it is added. If the reason for adding methanol to gasoline is to increase the volume of liquid fuel, this information implies that the MON rating of the blend will not be reduced.

The results of the single-cylinder engine test program indicate that adding 10% methanol to unleaded gasoline does not have a substantial effect on specific emissions or energy fuel economy. These results are based on equivalent operating conditions for the clear gasoline and the 10% methanol blend. The fact that the addition of methanol to gasoline alters the energy content of the fuel is apparent only when fuel consumption is measured on a volume or mass basis. From an emissions and energy consumption point of view, there are no substantial advantages or disadvantages in using up to 10% methanol blended with gasoline as fuel for spark ignition engines. In effect, methanol can be essentially a fuel extender.

The results of the chassis dynamometer study, when taken with the results of other investigators (*11, 12*), lead to the following conclusions. The addition of 10% methanol to unleaded gasoline will reduce the CO emissions of recent model vehicles that are not equipped with catalytic converters by approximately 30%. The addition of 10% methanol to unleaded gasoline does not substantially change the HC or NO_x emissions of late-model, noncatalyst vehicles. The miles traveled per gallon of fuel is less for the methanol blends than for unmodified gasoline. However, there is a slight increase in the miles traveled per million BTU's, indicating a slight increase in the energy efficiency in the unmodified engine when operating on the 10% methanol blend. Virtually all of these results can be attributed to the change in the fuel stoichiometry produced by the addition of methanol. The results for the stratified-charge engine vehicle also indicate that emissions and fuel economy are not substantially altered for this type of vehicle when using 10% methanol/gasoline blends.

Vehicle driveability is affected by adding methanol to the fuel. The stratified-charge engine vehicle appears to be somewhat more tolerant of the changes in fuel stoichiometry than conventional spark-ignition engine vehicles. Generally, the worst driveability problems are associated with start up and cold operation. Driveability problems will vary from vehicle to vehicle. Some drivers will notice little difference in the performance of their vehicle once it is fully warmed up. For others, the driveability problem may be severe with methanol/gasoline blends.

Generally speaking, most vehicles in the existing fleet could operate on blends of up to 10% methanol in gasoline without severe driveability problems or degradation of emissions and fuel economy performance. Some vehicles would require minor adjustments to provide reasonable performance and driveability.

Two major problems that have not been addressed in this investigation are the phase separation problems associated with methanol/gasoline blends contaminated with small amounts of water and the corrosion problems associated with methanol and the materials normally used in automotive fuel systems. For existing vehicles no simple economic solutions to these problems are evident. If methanol/gasoline blends are used, occasional phase separation and corrosion problems will arise which will cause inconvenience and additional expense to vehicle owners. Future vehicles could be designed with materials that would minimize or eliminate the corrosion problem, and at least one novel system (*15*) to capitalize on the phase separation problem is under investigation. Hence, if the need is strong enough, the corrosion and phase separation problems can be solved.

The key words to the future of methanol/gasoline blends for automotive use are need and availability. Technically, the operation of methanol/gasoline blends in automotive engines is feasible with some associated problems. Economically, methanol is not yet competitive with gasoline produced from petroleum, hence the need has not been strongly established. Since the need or market is not established, the capital expenses involved in producing methanol from coal or garbage are not presently justified. However, if 40–50% of the crude oil used in the U.S. to produce petroleum products should suddenly become unavailable, the need would be very real. The necessity of compete evaluation of methanol and other alternate fuels is evident.

Literature Cited

1. Bolt, J. A., "A Survey of Alcohol as a Motor Fuel," *Soc. Automot. Eng. Spec. Publ.* (June 1964) **254**.
2. Reed, T. B. et al., "Improved Performance of Internal Combustion Engines Using 5–30% Methanol in Gasoline," *Intersoc. Energy Convers. Conf. Proc. 9th* (Aug. 1974), **749104**.
3. Lee, W., Geffers, W., "Engine Performance Characteristics of Spark-Ignition Engines Burning Methanol and Methanol–Gasoline Mixtures," AIChE Meeting, Boston, Mass., Sept. 9, 1975.
4. Ingamells, J. C., Lindquist, R. J., "Methanol–Gasoline Blends—A Fuel Supplier's Viewpoint," in "Methanol as an Alternate Fuel," Vol. II, Engineering Foundation Conference, Henniker, N. H., July 1974.
5. "Alcohols, A Technical Assessment of Their Application as Fuels," *API Publ.* (June 1976) **4261**.
6. Johnson, R. T., Riley, R. K., "Single Cylinder Spark Ignition Engine Study of the Octane, Emissions, and Fuel Economy Characteristics of Methanol–Gasoline Blends," *SAE Tech. Pap.* (Feb. 1976) **760377**.
7. *Fed. Regist.* (Wednesday, Nov. 15, 1972) **37** (221) Part II.
8. "ASTM Manual for Rating Motor, Diesel, and Aviation Fuels," 1973–74 ed., American Society for Testing and Materials, Philadelphia, Pa.
9. Johnson, R. T., Riley, R. K., "Evaluation of Methyl Alcohol as a Vehicle Fuel Extender," Final Report for U.S. Department of Transportation contract: DOT-OS-40104, 1975, Report No. **DOT-TST-76-50**, NTIS No. PB251108/AS.
10. Stivender, D. L., "Development of a Fuel Based Mass Emission Measurement Procedure," *SAE Tech. Pap.* (June 1971) **710604**.
11. Wigg, E. E., Lung, R. S., "Methanol as a Gasoline Extender—Fuel Economy, Emissions, and High Temperature Driveability," *SAE Tech. Pap.* (Oct. 1974) **741008**.
12. Brinkman, N. D., Gallopoulos, N. E., Jackson, M. W., "Exhaust Emissions, Fuel Economy, and Driveability of Vehicles Fueled with Alcohol–Gasoline Blends," *SAE Tech. Pap.* (Feb. 1975) **750120**.
13. Johnson, R. T., Riley, R. K., Dalen, M. D., "Performance of Methanol–Gasoline Blends in a Stratified Charge Engine Vehicle," *SAE Tech. Pap.* (June 1976) **760546**.
14. *Fed. Regist.*, Environmental Protection Agency, "New Motor Vehicles and New Motor Vehicle Engines," (Nov. 15, 1972) **37** (221), Part II.
15. "Fuel System Allows Use of Methanol," *Automot. News* (April 26, 1976).

RECEIVED December 29, 1976.

Hydrogen Enrichment for Low-Emission Jet Combustion

RICHARD M. CLAYTON

Jet Propulsion Laboratory, California Institute of Technology,
Pasadena, CA 91103

Simultaneous gaseous pollutant emission indexes (g pollutant/kg fuel) for a research combustor with inlet air at 12.09 $\times$ 10^5 N/m^2 (11.9 atm) pressure and 727 K (849°F) temperature are as low as 1.0 for NO_x and CO and 0.5 for unburned HC. Emissions data are presented for hydrogen/ jet fuel (JP-5) mixes and for jet fuel only for premixed equivalence ratios from lean blowout to 0.65. Minimized emissions were achieved at an equivalence ratio of 0.38 using 10–12 mass % hydrogen in the total fuel to depress the lean blowout limit. They were not achievable with jet fuel alone because of the onset of lean blowout at an equivalence ratio too high to reduce the NO_x emission sufficiently.

The reduction of atmospheric pollutant emissions from transportation combustion sources has been a national concern for a number of years. Recently, the aviation community has directed its attention to the problem of reducing such emissions from aircraft gas turbine engines. The technological complexity of the problem is suggested in Ref. *1*, which lists 113 titles on the subject in the year 1973 alone. Refs. *2–6* contain later summaries of technological status and recent advances.

From a regulatory standpoint, the Environmental Protection Agency (EPA) has established standards (*7*) for aircraft jet engines that, by 1981, will require as much as 65% reduction in emissions of oxides of nitrogen (NO_x) for newly certified engines under takeoff power. Even greater levels of reduction of carbon monoxide (CO) and unburned hydrocarbons (HC) emissions will be required for ground-idle conditions. Moreover, concern over ozone depletion in the stratosphere from NO_x may ultimately lead to very stringent standards for high-altitiude cruise operation. And even though it is understood that the NO_x emissions result from high combustion temperatures under high-power operation and that CO and HC emissions result from reduced combustion efficiency under low-power operation, conventional approaches to combustor design changes generally fail to show promise of achieving the emission standards across the engine operating range. This lack of success is inherent in the opposing combustion requirements involved in reducing the two classes of emissions. Thus, the development of new combustor design concepts is required before combustion temperature can be reduced substantially without sacrificing flame stability and combustion efficiency.

The control over peak flame temperature that is available with premixed (including prevaporized liquid fuel by definition) combustion is well known, and this mode of combustion is presumed to be necessary to greater or lesser degree in all advanced combustor concepts for minimizing NO_x production. Several recent investigations have shown the potential of premixed, fuel-lean combustion in laboratory burners for achieving ultralow NO_x emissions (8, 9), but the leanness required at the higher power inlet air conditions is seriously close to the lean burning limit of hydrocarbon and air. Thus, the attractiveness of the premixed, lean-burning concept is tempered by the narrow margin between the mixture strength and combustor dwell times required for NO_x control and those required for flame stability, ignition, and high combustion efficiency.

The subject of this chapter is the concept of hydrogen enrichment for widening this combustion margin. The unique lean-burning properties of molecular hydrogen are used to depress the lean flammability limit of mixtures of premixed hydrogen, jet fuel, and air. The concept limit of 100% hydrogen premixed with air would permit burner operation down to idle power with no CO or HC emissions and high-power operations with ultralow levels of NO_x, but use of 100% hydrogen is not the essence of the concept. Instead, the object is to use as little hydrogen as possible to provide the necessary combustion margin. Effectively, the hydrogen replaces a portion of the ordinary jet fuel.

Hydrogen enrichment to reduce pollution emissions was originally introduced at this laboratory in 1973 for application to internal combustion engines, where the advantage of improved thermal efficiency is also available with lean burning. Many of the developments of the concept in application to automotive engines are reported in Refs. 10, 11, 12, and 13. The concept is also being applied to general aviation piston engines, where a significant extension in flight range for a fixed fuel load, as well as pollution reduction, is expected (14).

In the broadest view of the concept, the hydrogen might be provided by using a two-stage combustion system where a large portion of the total fuel flow to the system would be partially oxidized in the first stage in order to supply a hydrogen-bearing gas stream to the second stage. In the second stage the remainder of the system fuel would be burned lean after premixing with the remainder of the system air and the hydrogen-bearing product of the first stage.

However, application of the concept to continuous flow combustors has not reached the maturity of an integrated hydrogen-generation/combustor system, the present objective being to demonstrate the practical feasibility of the concept using bottled hydrogen. The essential proof of principle of the concept as applied to a laboratory pipe burner operating with premixed hydrogen/propane/air was obtained previously in the interesting experiments of Anderson (*15*).

Ultralow levels of emissions (levels considerably lower than EPA standards) have been adopted as targets. These targets are, in terms of emission index (g pollutant species/kg total fuel), 1.0 for NO_x and CO and 0.5 for HC.

The results presented here are for combustor inlet air conditions approximately equal to cruise power for a 30:1 compression ratio engine, nominally 12.09×10^5 N/m² (11.9 atm) at 727 K (849°F). Emissions data for hydrogen/jet fuel (JP-5) mixes and for jet fuel only were obtained over a range of equivalence ratios using a JPL research combustor designated Mod 2. The series of experiments from which the results are abstracted is described more fully in Ref. *16*.

The Hydrogen Enrichment Concept and Its Application to Jet Combustors

The basis of the hydrogen enrichment concept to provide low-temperature combustion stems from experimental observations that admixtures of hydrogen with HC/air mixtures so depress the lean flammability limit that burning can take place at ultralean combined fuel-to-air ratios. The potential reduction in lean limits provided by such "tertiary" mixtures can be illustrated by the use of Le Chatelier's formula which predicts the lean limit of any mixture of fuel gases from a knowledge of the lean limits for the individual fuel gases. This formula is (*17*):

$$L = \frac{100}{\dfrac{P_1}{N_1} + \dfrac{P_2}{N_2} + \cdots \dfrac{P_n}{N_n}} \tag{1}$$

where L = vol % of total fuel gas in a lean limit mixture with air; $P_1 \ldots {}_n$ = vol % of each combustible gas present in the fuel gas, calculated on an air- and inert-free basis so that $P_1 + P_2 + P_n = 100$; and

$N_1 \ldots _n$ = vol % of each combustible gas in a lean limit mixture of the individual fuel gas and air.

Equation 1 can be rearranged and expressed in terms of fuel/air mass ratios for more convenient use. The term mass refers to the mass flow rate ($\dot{M}$). For hydrogen/JP mixed fuels, Equation 1 then becomes:

$$R_{LM} = \frac{R_{LJP}}{1 + \beta_L F} \tag{2}$$

where R_{LM} = mass ratio of total fuel ($= \dot{M}H_2 + \dot{M}JP$) to air ($\dot{M}A$) in a lean limit mixture with air; F = mass fraction of hydrogen in the fuel mix ($= \dot{M}H_2/\dot{M}H_2 + \dot{M}JP$); R_{LJP} = mass ratio of JP fuel to air in a lean limit mixture with air; and $\beta_L = (R_{LPJ}/R_{LH2} - 1$; with R_{LH2} being analogous to R_{LJP}.

Furthermore, it can be shown that the stoichiometric fuel/air mass ratios for the hydrogen/JP mixed fuels can be expressed in the same form as Equation 2:

$$R_{SM} = \frac{R_{SJP}}{1 + \beta_S F} \tag{3}$$

where R_{SM} = mass ratio of total fuel to air in a stoichiometric mixture with air; R_{SJP} = mass ratio of JP fuel to air in a stoichiometric mixture with air; and $\beta_S = (R_{SJP}/R_{SH2}) - 1$, with R_{SH2} being analogous to R_{SJP}.

Dividing Equation 2 by Equation 3, substituting $\dot{M}H_2/(\dot{M}H_2 + \dot{M}JP)$ for F, and rearranging permits the expression of the system flow rate ratios required to satisfy particular lean limit equivalence ratios. Thus,

$$\left[\frac{\dot{M}JP}{\dot{M}H_2}\right]_{LM} = ER_{LM} \times R_{LJP} \left[\frac{\dot{M}A}{\dot{M}H_2}\right] - (1 + \beta_L) \tag{4}$$

Also, using the standard definition of operating equivalence ratio,

$$ER = \frac{\dot{M}H_2 + \dot{M}JP}{\dot{M}A \times R_{SM}} \tag{5}$$

it can be shown that the system flow rate ratios in general are related to ER as follows:

$$\left[\frac{\dot{M}JP}{\dot{M}H_2}\right] = ER \times R_{SJP} \left[\frac{\dot{M}A}{\dot{M}H_2}\right] - (1 + \beta_S) \tag{6}$$

Taking the molecular weight of JP-5 as 170 and its H/C ratio as 2.0 and using flammability limits for hydrogen and kerosene of 4.0% and 0.7% by volume (*17*), respectively, evaluation of Equation 6 yields the results plotted as solid lines on the operational map in Figure 1 for several selected air-to-hydrogen mass ratios. Evaluation of Equation 4 yields the solid curved line (ER_{LM}) in Figure 1, which is an estimate of the variation of lean flammability limit (a fundamental combustion property) with proportion of hydrogen in the fuel mix over a range of practical interest for low-temperature burning. This curve does not predict the lean operating limit line for a combustor using hydrogen-enriched fuel,

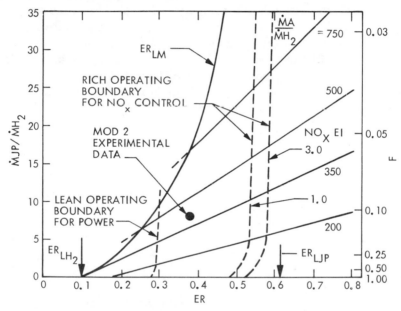

Figure 1.   Operating map for hydrogen enrichment concept, cruise power

which must be experimentally established. Nonetheless, the estimated flammability limit curve is an important reference because, with all other combustor factors fixed, fuels or fuel mixes with substantially lower flammability limits can be expected to provide substantially leaner operating limits.

The steady-state characteristics of practical significance for a lean-burning jet combustor intended to control pollutant emissions are its lean operating limits for efficient, stable combustion; i.e., the minimum operating equivalence ratios that yield acceptable levels of CO and unburned HC emissions. Those operating points that also yield acceptable

NO_x emission levels as well as sufficient heat release for turbine power requirements form the actual operating regime of overall acceptability.

The dashed lines in Figure 1 show conceptually rich and lean operational boundaries which, together with the flammability limit line, illustrate the potential benefits of using hydrogen enrichment for achieving acceptable lean-burning combustor operation. The illustration is for a typical subsonic-cruise power condition for a 30:1 compression ratio, high-bypass turbofan engine.

Two selected rich boundary lines are shown in Figure 1. One is based on the NO_x emission goal of the present work of 1.0 g NO_2/kg of total fuel, and the other is based on 3.0 g NO_2/kg of total fuel that has been suggested as a standard for cruise operation. The upper and lower end points of the boundary lines were evaluated for kinetically controlled NO_x production in one-dimensional flow with a 10.0-msec dwell time, using 100% jet fuel and 100% hydrogen, respectively. The data for intermediate mixed-fuel composition were interpolated for illustrative purposes. Both of these rich operating boundaries fall to the lean side of the flammability limit for jet fuel alone.

Less dwell time in the combustor, early quench techniques, and the use of other values of predicted lean flammability limits would now modify to some degree the placement of the rich boundary relative to the jet fuel lean limit. But the fact remains that the temperature reduction needed for ultralow NO_x emission levels requires that the heat release reaction take place at equivalence ratios approaching the limit for jet fuel/air combustion. Hence, at best, a low margin of acceptable flame stability, ignition, and combustion efficiency can be expected without some form of combustion aid. By virtue of its unique lean-burning qualities, molecular hydrogen substituted for a portion of the jet fuel can provide a significantly improved margin.

The lean operating boundary shown in Figure 1 (to its interception of the lean limit line) represents the line of constant turbine inlet gas temperature (1407 K or 2073°F) typically required for cruise power. This boundary line shows that there is a minimum allowable premixed equivalence ratio that satisfies power requirements. But, since air–film cooling and perhaps secondary air injection for temperature pattern factor adjustment (at the turbine inlet) will be required in an engine combustor, the useful lean boundary will lie possibly 20–30% to the right of that shown. Cooling requirements should be much reduced from current practice because of the ultralean (cooler) burning zone.

Thus, the conceptual, acceptable operating regime with hydrogen enrichment is enclosed by the lean and rich operating boundaries, the flammability limit line, and the abscissa which represents pure hydrogen fuel. The experimental data point shown in Figure 1 is discussed below and represents an operating condition demonstrated with the Mod 2 burner that satisfies all the ultralow emission goals. The position of the data point relative to the lean operating boundary would permit about 20% of the air mass flow to be used for cooling and/or dilution.

Experimental

Mod 2 Burner. The design parameters for the burner are summarized in Table I. Although the burner was not intended to be a scale version of the G.E. CF6-50 combustor, analogous design parameters for

Table I.   JPL Mod 2 Burner Design Specifications and Comparison with Typical Production Engine Burner[a] (16)

Specification Item	Mod 2 Burner[b]	Engine Burner[b]
Air total pressure	30.39×10^5 N/m² (30 atm)	30.39×10^5 N/m² (30 atm)
Air total temperature	812 K (1460 R)	821 K (1477 R)
Air flow rate	4.6 kg/sec (10 lbm/sec)	103.4 kg/sec (228 lbm/sec)
Chamber reference velocity	18.3 m/sec (60 ft/sec)	25.9 m/sec (85 ft/sec)
Chamber dwell time (no recirculation)	5.0 msec	2.6 msec
Chamber L/D (shape)	2.0 (cylindrical)	3.0 (annular)
Combustion length	36.8 cm (14.5 in)	34.8 cm (13.7 in.)
Combustion space rate[c]	0.88×10^6 J/hr-m³-N/m² (2.4×10^6 Btu/hr-ft³-atm)	2.2×10^6 J/hr-m³-N/m² (5.9×10^6 Btu/hr-ft³-atm)
Combustion equivalence ratio	LBO < ER < 1.0	> 1.0
Overall equivalence ratio	LBO < ER < 1.0	0.34
Air split for cooling	n.a.	~ 30%
Air split for dilution	n.a.	~ 38%
Premix reference velocity	157.8 m/sec (518 ft/sec)	n.a.
Premix Mach number	0.28	n.a. (0.27 at compressor discharge)
Premix dwell time	2.0 msec	n.a.
Premix length	30.5 cm (12 in.)	n.a.

[a] G.E. CF6-50. Data from Ref. 2.
[b] Takeoff conditions.
[c] At jet fuel $ER = 0.34$.

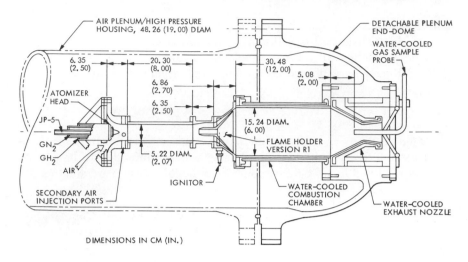

Figure 2.   Mod 2 burner, assembly, and test installation

a production version are shown for reference. The experimental burner has about 4% of the mass throughput and about 40% of the combustion space rate of the engine combustor.

For testing, the Mod 2 burner is housed within a heavily walled pressure vessel which also serves as a plenum chamber for the preheated inlet air supply. *See* Ref. *16* for a description of the test facility. The burner assembly is shown schematically in Figure 2 and is designed to use 100% of the air flow in the combustion process. Thus, air film cooling and air dilution which are normally used in an engine combustor are omitted. In this way, combustion effects from air injection are avoided for the concept evaluation. The cylindrical combustion chamber is water cooled, as are the sonic exhaust nozzle and gas sample probe.

The burner is intended to operate with a near-homogeneous fuel/air mixture. This is accomplished in the mixing section where gaseous hydrogen, finely atomized jet fuel, and the air are combined. The interfacial component between the combustion and premixing zones is a perforated conical flameholder which distributes the incoming high-velocity mixture across the chamber cross section. The jet flow produced by the perforations results in many small recirculation zones on the downstream side of the flameholder and thus provides flame stabilization. A spark ignitor on the downstream side of the flameholder provides ignition but is inactive once steady burning is obtained.

Fuel Injection/Premix Section. As is seen from Figure 2, premixing is accomplished via a coaxial flow scheme where the fuels are injected at the bell-mouthed entry to a 5.22-cm (2.07-in.) diameter straight cylindrical tube, 20.32 cm (8.00 in.) long. The flow area provides a space velocity of about 140 m/sec (460 ft/sec) for the cruise power condition. This results in a residence time in the premixed section of about 2.0 msec.

The fuel injector is a JPL-fabricated device designed to inject finely atomized jet fuel and gaseous hydrogen coaxially into the air stream. The injector assembly consists of a three-section feed manifold to which an atomizer head is attached as shown in Figure 2. A photograph of the atomizer is shown in Figure 3.

The intended operation of the atomizer head is depicted in Figure 4, where the flow from two of the several symmetrically spaced elements is schematically illustrated on a two-dimensional basis. Jet fuel atomization is accomplished pneumatically by injecting low-velocity streams of liquid fuel into high-velocity nitrogen flow. The nitrogen design flow was based on maintaining a mass ratio of hydrogen-to-jet fuel of unity or greater, with sonic jet velocity. In practice, a nearly constant nitrogen mass flow rate of 4–5% of the air mass flowrate for all operating conditions was used. Kinetic estimates of the effects of this additional nitrogen showed no tendency to increase the NO formation rate.

No attempt has been made to measure the drop sizes produced by this atomizer, but comparison of the design factors with atomization correlations for other pneumatic atomizers indicates that drop mean diameters in the range of 10–20 μm should be produced. Without allowance for further size reduction from evaporation during passage through

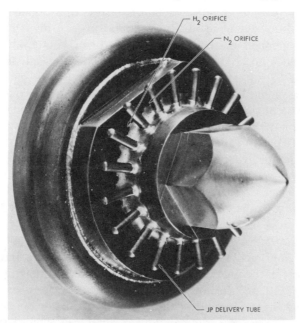

Figure 3.   JPL pneumatic atomizer head

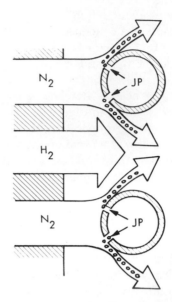

Figure 4. Schematic of atomizer operation

the mixer, this size range approaches that where the drops burn essentially as a vapor. Certainly this was the intent and justification for select-
ing that atomizing scheme.

As is depicted in Figure 4, the nitrogen/jet fuel flow pattern produced by the pairs of atomizer elements results in a region of high mass
flux midway between adjacent fuel delivery tubes. This region is formed
by the impingement of the bifurcated nitrogen flow from around the
tubes, and since nearly all the jet fuel must flow through that region, it
was decided to inject the hydrogen there. Thus, an axially directed,
subsonic hydrogen injection jet was placed at each of those locations.

The scope of these experiments did not permit a thorough evaluation
of the mixing performance of the injector/premixing section. But in
ancillary tests prior to the combustion runs wherein some information on
the distribution of hydrogen was obtained, it was clear that the hydrogen
tended to concentrate down the center of the mixing duct. This was
improved but not eliminated by adding four secondary air injection ports
near the throat plane of the bell-mouthed entry (Figure 2), and this
arrangement was used as the standard configuration.

Only one modification of the standard premixing section was used
during the exepriments. This consisted of doubling the number of air
injection ports in the plane of the existing ones and in moving the injector
~ 0.48 cm (0.19 in.) further into the entry section. As will be noted
below, this change apparently degraded the mixing.

Chamber Inlet and Flameholder Geometry. The design criteria
adopted for the perforated conical flameholder were to keep the mixture

velocity across the upstream side of the flameholder and through the flameholder flow area at least equal to the approach velocity from the mixing section and to distribute the mixture equally across the chamber area. The first requirement was approximated by sizing the annular flow passage between the dome wall and upstream surface of the flameholder in accordance with the progressive outflow from the flameholder perforations, and the flow dumped through the annular slit at the skirt edge of the cone. The second requirement was met by sizing and locating the drilled perforations to inject equal mixture flows into three equal annular areas of the chamber, with the remaining quarter of the flow dumped at the skirt. Flameholders were fabricated from AISI 310 stainless steel, 1.6 mm (0.062 in.) thick.

Several versions of this basic flameholder design have been used to improve its durability. This has been a persistent problem, especially for simulated power conditions greater than cruise (*16*). Suffice it to say that the bulk of the cruise results reported here were obtained with a version designated R1 which is shown in Figure 5. An alternate flameholder, version H (not shown), differed in mounting details and hole pattern in the apex region. However no clear difference in burner behavior was observed. Both had a total flow blockage of about 75% of the chamber cross sectional area.

The three overall configurations of the Mod 2 burner formed by combinations of the two mixing section and two flameholder versions mentioned above are sumamrized in Table II.

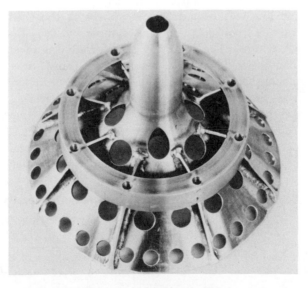

Figure 5. R1 flameholder

Table II.   Mod 2 Burner Variations Used for Cruise Power Experiments

Overall Configuration	Premix Section	Flameholder Version
AC	standard	H
C	standard	R1
BC	modified	R1

Combustion Gas Sample Probe and Sample Analysis. Combustion gas composition was analyzed on-line by means of a water-cooled probe, an electrically heated sample transfer line, and a so-called emission analysis bench. Gas samples were analyzed on a dry volumetric basis for O_2, CO_2, NO_x, CO, and unburned HC.

The water-cooled probe was mounted through the exhaust nozzle, as shown in Figure 2 on the centerline axis of the burner assembly, with the probe entrance located a fixed 36.8 cm (14.5 in.) from the flameholder skirt and in the subsonic approach flow to the nozzle sonic plane. Cooling provided by annular water passages was such that the gas sample stream temperature never exceeded 480 K (404°F) 1.2 m (4 ft) downstream of the probe entrance. The gas sample passage surface was stainless steel, plated with a gold film to discourage wall catalysis effects on NO_x. Passage diameter was 0.79 cm (0.31 in.) throughout.

The total sample from the probe was transferred to the emission bench location via heavily walled, stainless steel tubing about 91 m (300 ft) long, through the walls of which a low-voltage ac electrical current was passed to maintain approximately 423 K (302°F) wall temperature. The inside diameter of the transfer line was essentially the same as for the probe. The line was vented to atmosphere at the emission bench location after a small portion of the gas sample was withdrawn by the emission bench pumps. The total gas transferred was of the order of 1/2% of the combustor flow, which resulted in a transit time of 2 sec or less to the emission bench.

Gas analysis was conducted using chemiluminescence, FID, and NDIR instruments for NO_x, HC, and CO and CO_2, respectively. A paramagnetic instrument was used for O_2 concentration analysis.

Experimental Procedure. The emissions data were obtained during runs of 1–3 hr with nearly constant inlet air conditions simulating a cruise

Table III.   Plenum Air Conditions Used for Cruise Power Experiments

Air Flowrate Range [kg/sec (lbm/sec)]	Pressure [kN/m² (atm)]		Temperature [K (°F)]	
	Average	Standard Deviation	Average	Standard Deviation
1.52–1.83 (3.35–4.02)	1209 (11.93)	32 (0.32)	727 (849)	3 (5)

power level (*see* Table III). The burner was operated with jet fuel only and with hydrogen/jet fuel mixes over a range of input equivalence ratio, generally from a high of about 0.6 to a low just above total lean blowout. The latter was observed operationally as complete loss of flame. Because of the constant-area exhaust nozzle, it was necessary to modulate the air mass flow rate somewhat as the equivalence ratio was varied in order to maintain a nearly constant inlet air pressure in the plenum.

All experimental data were digitally recorded on magnetic tape and subsequently reduced by computer. Each data point was obtained with flow conditions held constant until continuously monitored gas analysis indicated steady values.

Results and Discussion

Results. Emission results are presented in terms of variables calculated from input (metered) flow rates. Emission index (EI) is the dependent variable with units of grams of pollutant species per kilogram of total fuel (hydrogen + jet fuel). Equivalence ratio (ER), on a total fuel-to-air basis, is used as the independent variable and is based on the stoichiometry of the particular hydrogen/jet fuel/air mixture. Mass % hydrogen in the total fuel is presented parametrically.

Figures 6, 7, and 8 show the results for CO, HC, and NO_x, respectively, for each of the three burner configurations, although there is no significant difference in the CO and HC emission trends for the three variations. The solid curves and open symbols represent data for jet fuel-only operation while the broken curves and solid symbols are for mixed fuel operation.

When the burner was operated with jet fuel only, equilibrium or lower levels of CO (Figure 6) and very low levels of HC (Figure 7) were observed for input ERs from 0.7 down to a little under 0.55, where both emissions increased rapidly with further decrease in ER. At an ER of about 0.53, a nearly discontinuous change to the sample gas composition occurred, which was most obvious in O_2 and CO_2 measurements. This change was ultimately interpreted as a partial blowout of the combustion process behind the flameholder, and this behavior is indicated as such in Figures 6 and 7. This transition was not apparent while operating the burner. Indeed, total lean blowout (LBO) was not observed until an ER of 0.37. Data for CO and HC in the transition region between partial and total LBO are identified with crosses in the open symbols. The interpretation of a partial blowout is discussed in Ref. *16*.

The dashed curves in Figures 6 and 7 are constructed through data for 10–12% input hydrogen in the total fuel. For clarity, those data are not numerically identified, but data for hydrogen concentrations falling outside this range are marked. Operation with mixed fuels eliminated the indication of partial blowout as deduced from O_2 and CO_2 measurements, and clearly, the presence of hydrogen provided a lower total LBO which, for 14% hydrogen, was observed at $ER = 0.26$. The improvement in flame stability with 10–12% hydrogen permitted operation of the burner down to an input ER of 0.38 before the CO emission index target of 1.0 g/kg fuel was exceeded, and, at that ER, a HC level of 0.1 g/kg fuel is still well below the target value of 0.5.

The premixed flame model and calculation procedures of Westenberg (*18*) were used to estimate kinetically limited CO oxidation for lean

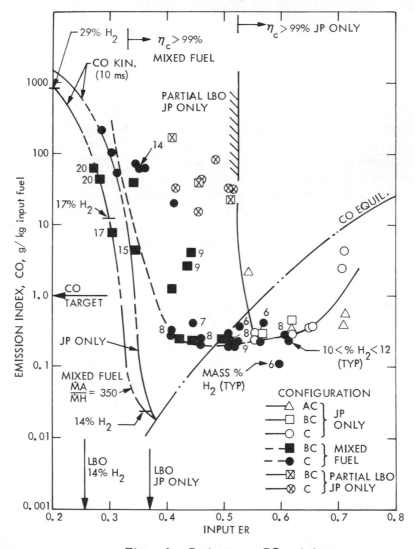

Figure 6.   Cruise power CO emissions

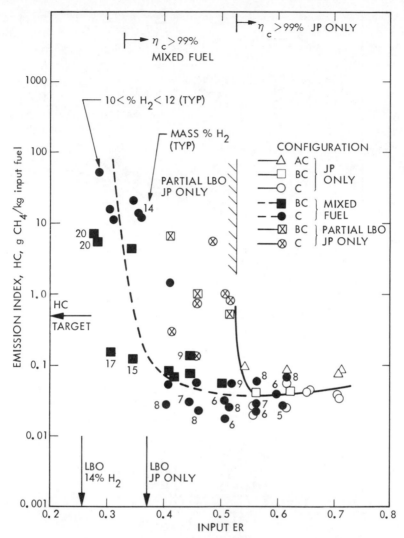

Figure 7.   Cruise power HC emissions

combustion at a constant 10-msec dwell time. The results of these calculations are shown in Figure 6 for jet fuel only and for mixed fuels at a
constant air-to-hydrogen mass ratio of 350. The effect of atomizer nitrogen on reducing flame temperature was accounted for. The proportion
of hydrogen in the total fuel varies continuously along the kinetic curve
for mixed fuels, as was illustrated in Figure 1, and three levels of hydrogen enrichment are identified on the mixed-fuel curve in Figure 6. These
are near the input hydrogen concentrations used in the experiments. For
simplicity, the kinetic curves are shown to intercept the equilibrium CO

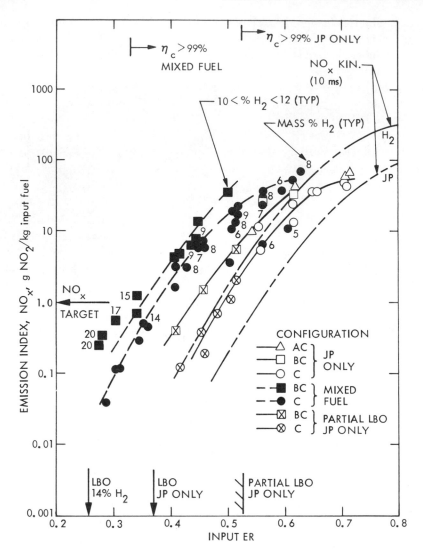

Figure 8.   Cruise power NO_x emissions

curve, but actually they approach the equilibrium line asymptotically in the fuel-rich direction. Also, the equilibrium curve is for jet fuel only (with atomizer nitrogen). This is acceptable because, when evaluated as a function of combined input *ER* and for the relatively low hydrogen concentrations of interest here, equilibrium CO changes only slightly for the mixed fuel.

Comparison of the experimental and calculated results shows that mixed-fuel operation yielded CO emissions approaching the kinetic esti-

mates, which is in significant contrast to the results for jet fuel-only operation. It is acknowledged that improvement in flame stabilization technique might improve lean-burning performance for jet fuel-only operation,
but then mixed-fuel performance would also undoubtedly be improved.
Thus, the feasibility of using mixed fuels to achieve a substantially de-
pressed lean operating limit in order to maintain ultralow CO and HC
emissions is well demonstrated by these results.

Contrary to the HC and CO results, NO_x emissions levels did change
for the different burner variations, as shown in Figure 8. This was especially true for variant BC, for which NO_x levels were higher, presumably
because of poorer mixing with the modified premixing section. No mixed-
fuels data were obtained for configuration AC because of the excessively
limited durability of flameholder version H. The open symbols with
crosses again represent data for *ER*s between partial and total LBO.

In Figure 8, the experimental NO_x results are compared with calculated results based on chemical kinetics for 100% jet fuel and
100% hydrogen, assuming one-dimensional flow for a constant 10-
msec dwell time. For the NO_x calculations shown in Figure 8, atomizer nitrogen was not included; therefore full flame temperature was
assumed. A computer program (*19*) was used to numerically integrate
the rate equations for the following set of elementary reversible reactions
generally accepted as important rate-controlling steps for post-flame
production of nitrogen oxides (*6*):

$$N + O_2 \rightleftarrows O + NO$$
$$O + N_2 \rightleftarrows N + NO$$
$$NO + M \rightleftarrows O + N + M$$
$$O_2 + M \rightleftarrows O + O + M$$
$$NO_2 + M \rightleftarrows O + NO + M$$
$$NO + O_2 \rightleftarrows O + NO_2$$
$$NO + NO \rightleftarrows N_2O + O$$
$$N_2O + M \rightleftarrows N_2 + O + M$$

The possibility of prompt NO was arbitrarily neglected since its
importance under lean conditions is likely to be small. These kinetic
estimates probably cannot be considered as predictions for the complex
recirculation region produced by the flameholder, but they do serve as
a reference for one limit of NO_x production.

Jet fuel-only operation produced NO_x trends which approximate
classical exponential dependency on *ER*, but with magnitudes greater
than the kinetics estimates. For a fixed input *ER*, introduction of hydro-

gen further increased the NO_x emission. This was expected because, at a fixed ER, flame temperature increases wtih hydrogen concentration in the total fuel. The kinetics estimates for 100% hydrogen and 100% jet fuel in Figure 2 show this. However, the increases observed in these experiments are much greater than the kinetics estimates suggest. This result is believed to be caused by certain premixing deficiencies which are discussed in Ref. *16*. Nonethless, configuration C yielded the target NO_x emission index of 1.0 g/kg fuel at an ER of 0.38 using 10–12% hydrogen.

Thus, configuration C exhibited an operating point for the cruise condition which satisfied all the target emissions simultaneously. Such an operating point was not achievable with jet fuel only, owing to excessive CO and HC at ER low enough to reduce NO_x to the target level.

A premixed primary zone ER of 0.38 at least approaches a feasible cruise design condition for the burner of an operating gas turbine engine of the 30:1 compression ratio class, where, typically, an overall ER of 0.31 is required for cruise power. A premixed heat release zone at 0.38 would leave about 20% of the total air flow for cooling and dilution. Cooling requirements should be much reduced from current practice because of the ultralean (cooler) burning zone. This cruise condition operating point was also discussed under "Hydrogen Enrichment," relative to the overall hydrogen-enrichment concept.

Total pressure loss from plenum to combustor exit at the end of cylindrical chamber was measured throughout the range of these experiments and varied with ER and flameholder version. The total range observed was from 5–8% of plenum pressure. For overall configuration C, the pressure loss was 7% at $ER = 0.38$. The quoted pressure losses include the losses at the bell-mouthed entry to the mixing section.

Conclusions

Conclusions based on these results are as follows:

(1) The ultralow target emission levels were simultaneously achieved under simulated cruise power conditions at a burning ER and with pressure losses amenable to practical combustor design. These levels were achievable with the Mod 2 burner with hydrogen enrichment but not with jet fuel only because of the onset of lean blowout at too high an ER to reduce the NO_x emission sufficiently.

(2) Premixing deficiencies, particularly with regard to the hydrogen, reduced the effectiveness of hydrogen enrichment for NO_x emission reduction and aggravated the flameholder durability problems in these experiments. Improved premixing should reduce the amount of hydrogen required.

(3) As reactant premixing is improved, the potential of lean burning for minimizing NO_x will be approached, but maintaining flame stabiliza-

tion and ultralow HC and CO emissions will become limiting practical considerations. Hydrogen enrichment can provide a significant lean-limit extension to minimize all emissions.

(4) Flame stabilization via a physical flameholder presents a difficult design/development problem because of the exposure of the component to the combustion zone. On the other hand, true achievement of the lowered combustion temperature potentially available with hydrogen enrichment should reduce the heat load to the exposed surfaces compared, for example, with present liner environments. Alternative flame stabilization schemes might also alleviate the problem.

(5) Effective implementation of hydrogen enrichment is not simple but is feasible with dedicated development. The bulk of the premixing problem has to be solved for any lean-burning scheme.

Literature Cited

1. Osgerby, I. T., "Literature Review of Turbine Combustor Modeling and Emissions," *AIAA J.* (1974) **12**(6), 743–754.
2. Rudey, R. A., "Status of Technological Advancement for Reducing Aircraft Gas Turbine Engine Pollutant Emissions," **NASA TMX-71846**, Lewis Research Center, Cleveland, Ohio, December 1975.
3. Niedzwiecki, R. W., "The Experimental Clean Combustor Program—Description and Status to November 1975," **NASA TMX-71849**, Lewis Research Center, Cleveland, Ohio, December 1975.
4. Yaffee, M. L., "NASA Seeks Clean Combustors by 1976," *Aviat. Week Space Technol.* (August 26, 1974).
5. Blazowski, W. S., Walsh, D. E., Mach, K. D., "Prediction of Aircraft Gas Turbine NO_x Emission Dependence on Engine Operating Parameters and Ambient Conditions," *AIAA/SAE Propul. Conf., 9th.* (November 5–7, 1973) **73-1275**.
6. Blazowski, W. S., "Aircraft Altitude Emissions: Fundamental Concepts and Future R&D Requirements," AIAA 1975 Aircraft Systems and Technology Meeting, Los Angeles, Calif., August 4–7, 1975, **75-1017**.
7. "Control of Air Pollution for Aircraft Engines—Emission Standards and Test Procedures for Aircraft," *Fed. Regist.* (July 17, 1973) **38**, 19088–19103.
8. Roffe, G., Ferri, A., "Effect of Premixing on Oxides of Nitrogen in Gas Turbine Combustors, **NASA CR-2657**, February 1976.
9. Anderson, D., "Effects of Equivalence Ratio and Dwell Time on Exhaust Emissions from an Experimental Prevaporizing Burner," **NASA TMX-71592**, Lewis Research Center, Cleveland Ohio, 1975.
10. Rupe, J. H., "System for Minimizing Internal Combustion Engine Pollution Emissions," U. S. Patent **3,906,913**, September 23, 1975.
11. Rupe, J. H., unpublished data (1973).
12. Hoehn, F. W., Baisley, R. L., Dowdy, M. W., "Advances in Ultralean Combustion Technology Using Hydrogen-Enriched Gasoline," *Intersoc. Energy Convers. Conf. Proc. 10th* (August 17–22, 1975) **759173**.
13. Houseman, J., Cerini, D., "On-Board Hydrogen Generator for a Partial Hydrogen Injection I. C. Engine," SAE National West Coast Meeting, Anaheim, Calif., August 12–16, 1974, **740600**.
14. Menard, W. A., Moynihan, P. I., Rupe, J. H., "New Potentials for Conventional Aircraft when Powered by Hydrogen-Enriched Gasoline," SAE Business Aircraft Meeting, Wichita, Kan., April 6–9, 1976, **760469**.

15. Anderson, D. N., "Effect of Hydrogen Injection on Stability and Emissions of an Experimental Premixed Prevaporized Propane Burner," **NASA TMX-3301**, Lewis Research Center, Cleveland Ohio, October 1975.
16. Clayton, R. M., "Reduction of Gaseous Pollutant Emissions from Gas Turbine Combustors Using Hydrogen-Enriched Jet Fuel—A Progress Report," *JPL Tech. Mem.* (October 1, 1976) **33-790**.
17. Coward, H. F., Jones, G. W., "Limits of Flammability of Gases and Vapors," *U.S. Bur. Mines Bull.* (1952) **503**.
18. Westenberg, A. A., "Kinetics of NO and CO in Lean, Premixed Hydrocarbon-Air Flames," *Combust. Sci. Technol.* (1971) (4) 59–64.
19. Laurendeau, N., Sawyer, R. F., "General Reaction Rate Problems: Combined Integration and Steady State Analysis," *Univ. Calif., Berkeley, Publ. Eng.* (December 1970) **TS-70-14**.

RECEIVED December 29, 1976. Research carried out under contract NAS7-100, sponsored by the National Aeronautics and Space Administration.

INDEX

The text of this book is set in 10 point Caledonia with two points of leading. The chapter numerals are set in 30 point Garamond; the chapter titles are set in 18 point Garamond Bold.

The book is printed offset on Text White Opaque 50-pound. The cover is Joanna Book Binding blue linen.

Jacket design by Alan Kahan
Editing and production by Virginia deHaven Orr.

The book was composed by Service Composition Co., Baltimore, Md., printed and bound by The Maple Press Co., York, Pa.